Practical Ideas That Really Work
for Teaching Math Problem Solving

Gail R. Ryser

James R. Patton

Edward A. Polloway

Kathleen McConnell

8700 Shoal Creek Boulevard
Austin, Texas 78757-6897
800/897-3202 Fax 800/397-7633
www.proedinc.com

© 2006 by PRO-ED, Inc.
8700 Shoal Creek Boulevard
Austin, Texas 78757-6897
800/897-3202   Fax 800/397-7633
www.proedinc.com

All rights reserved. No part of the material protected by this copyright notice may be reproduced or used in any form or by any means, electronic or mechanical, including photocopying, recording, or by any information storage and retrieval system, without prior written permission of the copyright owner.

NOTICE: PRO-ED grants permission to the user of this material to copy the forms for teaching purposes. Duplication of this material for commercial use is prohibited.

ISBN: 1-4164-0198-9

Printed in the United States of America

1 2 3 4 5 6 7 8 9 10   10 09 08 07 06

Contents

Introduction		1
Idea 1	4 Steps To Solve It	11
Idea 2	Step by Step	15
Idea 3	Graphic Organizers	17
Idea 4	Log and Checklist	31
Idea 5	Mnemonics	37
Idea 6	Everyday Math	43
Idea 7	Write It, Explain It	45
Idea 8	Get Real	49
Idea 9	Define the Word or Symbol	51
Idea 10	Vocabulary Games	57
Idea 11	Word Clues	63
Idea 12	Math Jeopardy	73
Idea 13	Highlighting	81
Idea 14	What's Missing?	83
Idea 15	What's the Question?	87
Idea 16	Can't Tell	91
Idea 17	Get Graphic	93
Idea 18	Use Manipulatives	97
Idea 19	Draw a Diagram	103
Idea 20	Find a Pattern	109
Idea 21	Make a Table	119
Idea 22	Make a List	127
Idea 23	Work Backward	135
Idea 24	Guess and Test	141
Idea 25	Estimation Games	149

Idea 26 The Price Is Right **163**

Idea 27 Beginning Problem Solving **165**

Idea 28 Which Operation? **179**

Idea 29 Formulas **183**

Idea 30 Maze It or Cloze It **189**

Idea 31 Peer Partners **195**

Idea 32 Assigning Students to Groups **199**

Idea 33 Collaborative Problem Solving **205**

Idea 34 Use a Matrix **211**

Introduction

Practical Ideas for Teaching Math Problem Solving was written to assist teachers in providing instructional activities and learning strategies that will promote effective problem solving in students who experience difficulties with problem-solving skills in school. The book builds on the feedback that we have collectively received about our prior books in the Practical Ideas series. We have learned that teachers find it useful to have a repertoire of appropriate instructional strategies to use.

Background

There is substantial documentation of the difficulties experienced by students in mathematics (Cawley, Parmar, Yan, & Miller, 1998; Gross-Tsur, Manor, & Shalev, 1996; Miller, Butler, & Lee, 1998). Although math difficulties are consistently found in students with learning disabilities, such problems are also commonly noted in the population of students in general (Miller & Mercer, 1997). This book focuses on teaching mathematical problem solving. The importance of mathematical problem solving is highlighted by the National Council of Teachers of Mathematics (NCTM; 2000). The NCTM developed 10 standards that describe the mathematical understanding, knowledge, and skills that students should acquire from Pre-K through Grade 12. Five standards describe the content, and five standards describe the processes that are important in mathematics. First of the five process standards is problem solving. NCTM takes the position that problem solving is an integral part of mathematics. By engaging in problem solving, students acquire habits of thinking and curiosity that carry over into the real world. The NCTM problem-solving standard is that instructional programs from prekindergarten through Grade 12 should enable all students to

- build new mathematical knowledge through problem solving,
- solve problems that arise in mathematics and in other contexts,
- apply and adapt a variety of appropriate strategies to solve problems, and
- monitor and reflect on the process of mathematical problem solving.

Within mathematics there has been a common approach to consider computation and problem solving as two major components. This somewhat artificial distinction parallels the similar way of conceptualizing reading instruction, which includes a focus on word recognition and reading comprehension. This book focuses on problem-solving strategies, but because computation and problem solving cannot be completely separated, a number of the strategies also have implications for instruction related to mathematics computation.

Components

Practical Ideas That Really Work for Teaching Math Problem Solving is intended for use with students who are using problem-solving skills that encompass addition and subtraction through prealgebra or algebra skills. The majority of the ideas are applicable for students who are working on Grades 3 through 8 mathematics skills, although some ideas are applicable for younger students. It includes the following two primary components.

- **Evaluation Form with a Rating Scale and Ideas Matrix.** The Evaluation Form has two features. The Rating Scale serves as a criterion-reference measure for evaluating problem-solving behaviors that impact student learning. The specific scale items provide descriptors that correlate to four key criteria relative to problem solving: reading the problem, deriving meaning from the problem, solving the problem, and looking back and building on success with problems. To complement the scale, the Ideas Matrix incorporated in the Evaluation Form offers a link between the results obtained from administration of the Rating Scale to possible classroom interventions. The intent of the Ideas Matrix is for teachers to select appropriate interventions that match the specific needs of an individual or group of students.
- **Practical Ideas Manual.** The core of this book consists of 34 ideas that can be effective in assisting students who are experiencing difficulties in math problem solving. The ideas were developed using scientifically validated practices, including reviews through research and validation through classroom practice. The ideas cover the range of primary areas identified within the Rating Scale and across the age levels for which this manual

is intended. Each of the ideas includes an explanation of the idea; strategies for implementing the idea; and reproducible worksheets, examples, and illustrations to assist the teacher in classroom use.

The Rating Scale

The Rating Scale in the Evaluation Form is a criterion-reference instrument. It was designed to be used by teachers to determine students' learning difficulties relative to problem solving in mathematics. The Rating Scale will assist teachers in their efforts to carefully and thoroughly assess specific problems across four major areas.

Item Development

Specific items within the Rating Scale are based on the mathematical problem-solving model developed by Polya (1957). The model identified four principles necessary for good mathematical problem solving. Polya believed that problem-solving skills could be taught, and our agreement with this tenet, provided a major impetus for developing this book.

Initially, a draft set of 20 items was developed for review. The manual's authors and several colleagues analyzed the items for their importance and content validity. After this review, the Rating Scale was formalized and now includes a total of 12 items. Three items describe each of the four criteria: (a) reads the problem, (b) derives meaning from the problem by using organizational tools and strategies, (c) solves the problem, and (d) looks back and builds on success with problems. Educators should rate the items on the scale using a 4-point Likert system, with 3 = *Consistently exhibits the behavior,* 2 = *Frequently exhibits the behavior,* 1 = *Sometimes exhibits the behavior,* and 0 = *Never or rarely exhibits the behavior.* For each of the four areas, the range of possible scores is 0 to 9; the lower the score, the more significant the students' mathematical problem-solving needs are.

Field-Testing the Rating Scale

The criterion-referenced measure was field tested in three school districts in Texas and three school districts in Virginia with 118 students. Of these students, 75 were identified as having difficulties with math problem solving; the other 43 did not have difficulties with math problem solving.

For the sample of students with math problem-solving difficulties, 33 had no disability, 6 had ADHD, 16 had a learning disability in math, 4 had a learning disability other than in math, 6 were English language learners, 1 had a speech or language impairment, 2 were classified with another disability, and 7 had more than one disability reported (including a learning disability in math). Thirty-three students were male, and 42 were female. Forty-seven were European American, 6 were African American, 19 were Hispanic American, 2 were Asian, and 1 student was more than one race/ethnicity. The students were in Grades 3 through 8. For the sample of students without math problem-solving difficulties, 21 were male, and 22 were female. Twenty-nine were European American, 5 were African American, and 9 were Hispanic American. The students were in Grades 3 through 8.

An item analysis was conducted, and the resulting reliability coefficient was .96. In addition, we compared the mean ratings for the two groups—students identified as having math difficulties and students not identified—using a t ratio. Our hypothesis was that students identified as having math difficulties would be rated lower than students not identified. The mean for students identified as having math difficulties was 12, and the mean for students not identified was 28. The mean difference between the two groups was large enough to support our hypothesis. The probability was < .001. We can conclude that the Rating Scale is sensitive enough to discriminate between the two groups.

The Manual

As with the other books in the Practical Ideas series, this manual was built on our discussions and consultations with general and special education classroom teachers. These individuals have consistently emphasized the importance of having access to materials that are practical, that are relatively easy to implement in either inclusive or specialized classrooms, and that do not require a significant amount of time to create learning activities. These basic principles underscored the development of the manual.

At the same time, we are also cognizant of the importance of presenting strategies, to the maximum extent possible, that are scientifically validated. Consistent with the basic tenets of the No Child Left Behind Act of 2001 and the most recent authorization of the Individuals with Disabilities Education Act (2004), we have thus complemented our emphasis on practical and easy-to-use strategies with an emphasis on those that have been validated. We have been influenced by the system developed by Patton (1994), in which he identified strategies according to the following categories: empirically (or scientifically) validated, literature based, and field based. We have used this concept in selecting appropriate activities for the examples in this manual. The result is a book with 34 ideas, most with reproducible blackline masters, and all grounded in our research and collective experiences, as well as those of the many educators who advised us and shared information with us.

In special education, our emphasis on assessment unfortunately has not always been matched with a similar emphasis on effective intervention, and thus this manual focuses on the use of assessment data in designing effective education interventions. The combination of the Rating Scale and Ideas Matrix facilitate the use of the manual. We designed the Ideas Matrix so that educators can make the direct link between the information provided by the Rating Scale and instruction in the classroom. We believe that this format stays true to our purpose of presenting information that is practical and useful.

Directions for Using the Materials

Step 1: Collect Student Information

The first step is to complete the first page of the Evaluation Form for the child who is identified as having difficulties with math problem solving. As an example, a completed Evaluation Form is provided in Figure 1. Space is provided on the front of the form for pertinent information about the student being rated, including name, birth date, age, school, grade, rater, and subject area. In addition, the dates the student is observed and the amount of time the rater spends with the student can be recorded here. Also included on the front of the form is the NCTM process standard for problem solving.

Step 2: Rate the Behaviors of the Student

The second page of the Evaluation Form contains the Rating Scale. The scale items are divided into the four criterion areas previously noted. Instructions for administering and scoring the items are provided on the form. Spaces are provided to total the items for each criterion, to check the area to target for immediate intervention, and to record the intervention idea and its starting date.

Step 3: Choose the Ideas To Implement

The third page of the Evaluation Form contains the Ideas Matrix. After choosing the areas to target for immediate intervention, the professional should turn to the Ideas Matrix and select one or more interventions that correspond to that problem. The professional should write the idea number and the starting date on the space provided on the rating scale.

For example, in Figure 1, Samuel received the lowest ratings in the area "Derives meaning from the problem by using organizational tools and strategies." His teacher has targeted this area and has chosen Ideas 1, 2, and 19 from the Ideas Matrix.

Step 4: Read and Review the Practical Ideas That Have Been Selected

Within the manual, the ideas are discussed in terms of their intent and implementation. After selecting the idea that is matched to the needs of the student, the idea can then be planned for implementation. These individual ideas should be integrated into an overall instructional design and should be reflected in classroom lessons that focus on the particular learning objective.

Although the Rating Scale and the Ideas Matrix might convey the sense that this approach is intended for use on a clinical, one-to-one basis, the information can be effectively used with larger groups of students. In fact, many of the most effective ideas that are designed for students with learning difficulties work very effectively with many students who often are not given the benefit of strategy training. These ideas can be implemented with the entire class, which will eliminate the need to create separate ideas for only some students.

Step 5: Evaluation

After implementation, teachers should complete the assessment cycle by evaluating the results of the intervention strategy. By following a model that (a) begins with the assessment of need, (b) leads to the development of an instructional plan, (c) follows with the implementation plan, and (d) concludes with the evaluation of its effectiveness, teachers can ensure a responsive educational program that enables students to enhance their achievement in the area of problem solving. The information within this manual can be correlated with annual goal setting and unit planning done for students in mathematics.

Page 4 of the Evaluation Form includes an Intervention Plan. This form can be used to assist in the evaluation of the idea and to document other assessment information. In Figure 1, Samuel had a score on the *Comprehensive Mathematical Abilities Test* (CMAT) of 91. In addition, his teacher has noted that he does not know how to begin solving problems and seldom uses strategies to solve problems. He has a grade point average of 32. His teacher will use Ideas 1 and 2 as whole-class activities and will set up a center with specific strategies beginning with Idea 19, Draw a Diagram.

References

Cawley, J. F., Fitzmaurice-Hayes, A. M., & Shaw, R. A. (1988). *Mathematics for the mildly handicapped: A guide to curriculum and instruction.* Boston: Allyn & Bacon.

Cawley, J. F., Parmar, R. S., Yan, W. F., & Miller, J. H. (1998). Arithmetic computation performance of students with learning disabilities: Implications for curriculum. *Learning Disabilities Research & Practice, 13,* 68–74.

Fuchs, L. S., Fuchs, D., Hamlett, C. L., & Appleton, A. C. (2003). Explicitly teaching for transfer: Effects on the mathematical problem-solving performance of students with mathematics disabilities. *Learning Disabilities Research & Practice, 17,* 90–106.

Gross-Tsur, V., Manor, O., & Shalev, R. S. (1996). Developmental dyscalculia: Prevalence and demographic features. *Developmental Medicine and Child Neurology, 38,* 25–33.

Higgins, J., McConnell, K., Patton, J. R., & Ryser, G. R. (2003). *Practical ideas that really work for students with dyslexia and other reading disorders.* Austin, TX: PRO-ED.

Jitendra, A. K. (2002). Teaching students math problem solving through graphic representations. *Teaching Exceptional Children, 34*(4), 34–38.

Jitendra, A. K., Hoff, K., & Beck, M. (1999). Teaching middle school students with learning disabilities to solve multistep word problems using a schema-based approach. *Remedial and Special Education, 20*(1), 50–64.

Maccini, P., & Hughes, C. A. (2000). Effects of a problem-solving strategy on the introductory algebra performance of secondary students with learning disabilities. *Learning Disabilities Research & Practice, 15,* 10–21.

McConnell, K., & Ryser, G. R. (2005). *Practical ideas that really work for students with ADHD: Grade 5 through grade 12* (2nd ed.). Austin, TX: PRO-ED.

McConnell, K., Campos, D., & Ryser, G. R. (2006). *Practical ideas that really work for English language learners.* Austin, TX: PRO-ED.

Miller, S. P., Butler, F. M., & Lee, K. (1998). Validated practices for teaching mathematics to students with learning disabilities: A review of the literature. *Focus on Exceptional Children, 31*(1), 1–24.

Miller, S. P., & Mercer, C. D. (1993). Mnemonics: Enhancing the math performance of students with learning difficulties. *Intervention in School and Clinic, 29,* 47–56.

Miller, S. P., & Mercer, C. D. (1997). Educational aspects of mathematics disabilities. *Journal of Learning Disabilities, 30,* 47–56.

National Council of Teachers of Mathematics. (2000). *Principles and standards for school mathematics.* Reston, VA: Author.

Patton, J. R. (1994). Practical recommendations for using homework with students who are learning disabled. *Journal of Learning Disabilities, 27,* 570–578.

Polya, G. (1957). *How to solve it: A new aspect of mathematical method* (2nd ed.). Princeton, NJ: Princeton University Press.

Ryser, G. R., & McConnell, K. (2003). *Practical ideas that really work for students who are gifted.* Austin, TX: PRO-ED.

Stein, M., Kinder, D., Silbert, J., & Carnine, D. W. (2006). *Designing effective mathematics instruction: A direct instruction approach* (4th ed.). Upper Saddle River, NJ: Prentice Hall.

Resources for Ideas

Bilsky, L. H., Blachman, S., Chi, C., Mui, A. C., & Winter, P. (1986). Comprehension strategies in math problem and story contexts. *Cognition and Instruction, 3,* 109–126.
Ideas 14, 16

Blachowicz, C., & Fisher, P. (2000). Teaching vocabulary. In M. Kamil, P. Mosenthal, P. D. Pearson, & R. Barr (Eds.), *Handbook of reading research* (Vol. 3, pp. 503–523). Mahwah, NJ: Erlbaum.
Ideas 9, 10, 11, 12

Braselton, S., & Decker, B. G. (1994). Using graphic organizers to improve the reading of mathematics. *The Reading Teacher, 48*(3), 276–281.
Ideas 2, 3, 17

Cass, M., Cates, D., Smith, M., & Jackson, C. (2003). Effects of manipulative instruction on solving area and perimeter problems by students with learning disabilities. *Learning Disabilities Research & Practice, 18,* 112–120.
Idea 18

Cawley, J. F., & Parmar, R. S. (1994). Structuring word problems for diagnostic teaching: Helping teachers meet the needs of students with mild disabilities. *Teaching Exceptional Children, 26*(4), 16–21.
Idea 34

Darch, C., Carnine, D., & Gersten, R. (1984). Explicit instruction in mathematics problem solving. *Journal of Educational Research, 77,* 351–359.
Idea 28

Friel, S. N., Curcio, F. R., & Bright, G. W. (2001). Making sense of graphs: Critical factors influencing comprehension and instructional implications. *Journal for Research in Mathematics Education, 32*(2), 124–158.
Idea 17

Fuchs, L. S., Fuchs, D., Hamlett, C. L., & Appleton, A. C. (2003). Explicitly teaching for transfer: Effects on the mathematical problem-solving performance of students with mathematics disabilities. *Learning Disabilities Research & Practice, 17,* 90–106.
Idea 34

Gick, M. L., & Holyoak, K. J. (1983). Schema induction and analogical transfer. *Cognitive Psychology, 15,* 1–38.
Ideas 20, 21, 22, 23, 24, 29

Goldin, G. A. (2000). Affective pathways and representation in mathematical problem solving. *Mathematical Thinking and Learning, 2,* 209–219.
Ideas 6, 7, 8, 15

Hegarty, M., & Kozhevnikov, M. (1999). Types of visual-spatial representations and mathematical problem solving. *Journal of Educational Psychology, 91*(4), 684–689.
Ideas 3, 17, 19

Hohn, R. L., & Frey, B. (2002). Heuristic training and performance in elementary mathematical problem solving. *Journal of Educational Research, 95,* 374–380.
Idea 5

Ives, B., & Hoy, C. (2003). Graphic organizers applied to higher-level secondary mathematics. *Learning Disabilities Research & Practice, 18*(1), 36–51.
Ideas 3, 17

Jitendra, A. (2002). Teaching students math problem-solving through graphic representations. *Teaching Exceptional Children, 34*(4), 34–38.
Ideas 27, 28

Kamii, C., Rummelsburg, J., & Kari, A. (2005). Teaching arithmetic to low-performing, low-SES first graders. *Journal of Mathematical Behavior, 24*(1), 39–50.
Ideas 10, 12, 25, 26

Leonard, J. (2001). How group composition influenced the achievement of sixth-grade mathematics students. *Mathematical Thinking and Learning, 3*(2/3), 175–200.
Ideas 31, 32, 33

Lesh, R., & Harel, G. (2003). Problem solving, modeling, and local conceptual development. *Mathematical Thinking and Learning, 5*(2/3), 157–189.
Ideas 1, 2, 3, 4, 5, 6, 7, 8, 20, 21, 22, 23, 24

Maccini, P. (1998). *Effects of an instructional strategy incorporating concrete problem representation of the introductory algebra performance of secondary students with learning disabilities.* Unpublished doctoral dissertation, Pennsylvania State University, University Park, PA.
Idea 18

Maccini, P., & Gagnon, J. C. (2000). Best practices for teaching mathematics to secondary students with special needs. *Focus on Exceptional Children, 32*(5), 1–24.
Ideas 5, 8, 29, 31

Maccini, P., & Hughes, C. A. (2000). Effects of a problem-solving strategy on the introductory algebra performance of secondary students with learning disabilities. *Learning Disabilities Research & Practice, 15,* 10–21.
Idea 18

Maccini, P., & Ruhl, K. L. (2000). Effects of a graduated instructional sequence on the algebraic subtraction of integers by secondary students with learning disabilities. *Education and Treatment of Children, 23,* 468–489.
Ideas 1, 5, 18, 19

Marsh, L. G., & Cooke, N. L. (1996). The effects if using manipulatives in teaching math problem solving to students with learning disabilities. *Learning Disabilities Research & Practice, 11*(1), 58–65.
Idea 18

Mercer, C. D., & Miller, S. P. (1992). Teaching students with learning problems in math to acquire, understand, and apply basic math facts. *Remedial and Special Education, 13*(3), 19–35, 61.
Ideas 1, 5, 18, 19

Miller, S. P., & Mercer, C. D. (1997). Educational aspects of mathematics disabilities. *Journal of Learning Disabilities, 30,* 47–56.
Ideas 1, 2, 18

Miller, S. P., & Mercer, C. D. (1993). Using data to learn about concrete-semiconcrete-abstract instruction for students with mild learning disabilities. *Learning Disabilities Research & Practice, 8,* 89–96.
Idea 18

Miller, S. P., Mercer, C. D., & Dillon, A. S. (1992). CSA: Acquiring and retaining math skills. *Intervention in School and Clinic, 28,* 105–110.
Idea 27

Miller, S. P., Strawser, S., & Mercer, C. D. (1996). Promoting strategic math performance among students with learning disabilities. *LD Forum, 21*(2), 34–40.
Idea 27

Montague, M., & Bos, C. S. (1986). The effect of cognitive strategy training in verbal math problem solving performance of learning disabled adolescents. *Journal of Learning Disabilities, 19,* 26–33.
Ideas 1, 2, 3, 4, 5, 13, 19, 25, 26

Naglieri, J. A., & Gottling, S. H. (1995). A study of planning and mathematics instruction for students with learning disabilities. *Psychological Reports, 76,* 1343–1354.
Ideas 1, 4, 7

Naglieri, J. A., & Gottling, S. H. (1997). Mathematics instruction and PASS cognitive processes: An intervention study. *Journal of Learning Disabilities, 30,* 513–520.
Ideas 1, 4, 7

Parmar, R. S., Cawley, J. F., & Frazita, R. R. (1996). Word problem-solving by students with and without mild disabilities. *Exceptional Children, 62,* 415–429.
Ideas 13, 15, 16, 27, 28, 34

Ross, J. A. (1995). Effects of feedback on student behavior in cooperative learning groups in a grade 7 math class. *Elementary School Journal, 96*(2), 125–144.
Ideas 31, 32, 33

Scott, J. A., Jamieson-Noel, D., & Asselin, M. (2003). Vocabulary instruction throughout the day in twenty-three Canadian upper-elementary classrooms. *The Elementary School Journal, 103*(3), 269–286.
Ideas 9, 10, 11

Swanson, H. L., & Beebe-Frankenberger, M. (2004). The relationship between working memory and mathematical problem solving in children at risk and not at risk for serious math difficulties. *Journal of Educational Psychology, 96,* 471–491.
Idea 29

Topping, K. J., Campbell, J., Douglas, W., & Smith, A. (2003). Cross-age tutoring in mathematics with seven- and 11-year-olds: Influence on mathematical vocabulary, strategic dialogue and self-concept. *Educational Research, 45*(3), 287–308.
Ideas 10, 12, 25, 26

Van Garderen, D., & Montague, M. (2003). Visual-spatial representation, mathematical problem solving, and students of varying abilities. *Learning Disabilities Research & Pratice, 18,* 246–254.
Idea 19

Verschaffel, L., De Corte, E., Lasure, S., Van Vaerenbergh, G., Bogaerts, H., & Ratinckx, E. (1999). Learning to solve mathematical application problems: A design experiment with fifth graders. *Mathematical Thinking and Learning, 1,* 195–229.
Ideas 1, 2, 3, 4, 5, 6, 8, 19, 20, 21, 22, 23, 24, 29

Watson, J. M., & Chick, H. L. (2001). Factors influencing the outcomes of collaborative mathematical problem solving: An introduction. *Mathematical Thinking and Learning, 3,* 125–173.
Ideas 31, 32, 33

Witzel, B. S. (2005). Using CRA to teach algebra to students with math difficulties in inclusive settings. *Learning Disabilities: A Contemporary Journal, 3,* 49–60.
Idea 18

Practical Ideas That Really Work for Teaching Math Problem Solving

Gail R. Ryser • James R. Patton
Edward A. Polloway • Kathleen McConnell

Evaluation Form

Name Samuel Dixon

Birth Date June 8, 1994   **Age** 12

School Mt Pleasant Middle School   **Grade** 6

Rater Ms. Danielle Johnson

Subject Area Mathematics

Dates Student Observed: From Aug. 10 **To** Oct. 10

Amount of Time Spent with Student:

Per Day 55 min.   **Per Week** 5 days

National Council of Teachers of Mathematics Process Standard: Problem Solving

Process Standard

Instructional programs from prekindergarten through Grade 12 should enable all students to

- build new mathematical knowledge through problem solving,
- solve problems that arise in mathematics and in other contexts,
- apply and adapt a variety of appropriate strategies to solve problems, and
- monitor and reflect on the process of mathematical problem solving.

Goal

Problem solving means engaging in a task for which the solution method is not known in advance. In order to find a solution, students must draw on their knowledge, and through this process, they will often develop new mathematical understandings. Solving problems is not only a goal of learning mathematics but also a major means of doing so. Students should have frequent opportunities to formulate, grapple with, and solve complex problems that require a significant amount of effort and should then be encouraged to reflect on their thinking.

By learning problem solving in mathematics, students should acquire ways of thinking, habits of persistence and curiosity, and confidence in unfamiliar situations that will serve them well outside the mathematics classroom. In everyday life and in the workplace, being a good problem solver can lead to great advantages.

Note. From *Principles and Standards for School Mathematics* (p. 52), by National Council of Teachers of Mathematics, 2000, Reston, VA: National Council of Teachers of Mathematics. Copyright 2000 by National Council of Teachers of Mathematics. Standards are listed with the permission of the National Council of Teachers of Mathematics (NCTM). NCTM does not endorse the content or validity of these alignments.

Figure 1. Sample Evaluation Form, filled out for Samuel.

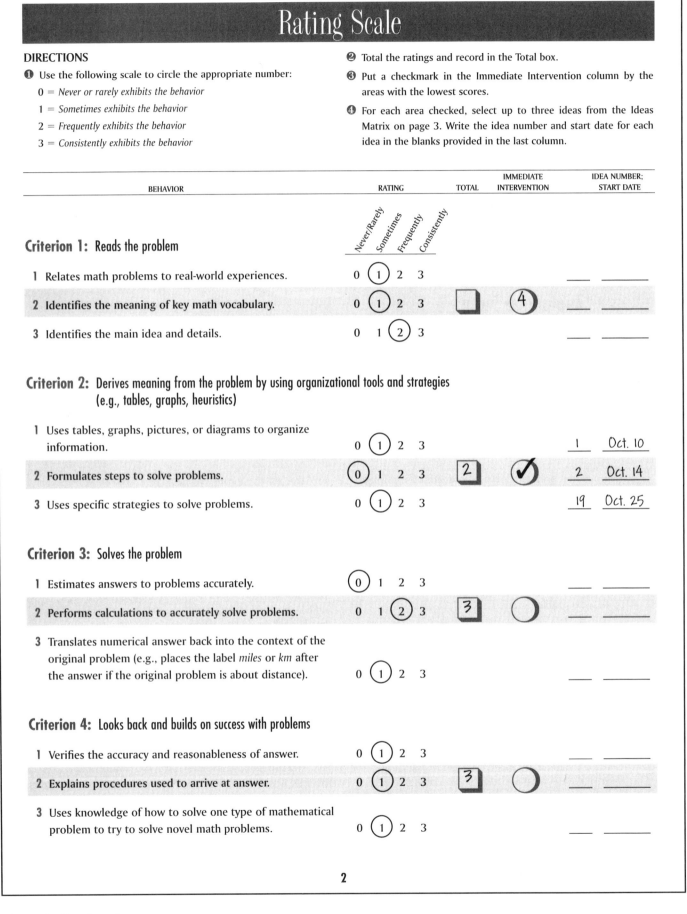

Figure 1. Continued.

Ideas Matrix

Ideas	Criterion 1 Reads Problem			Criterion 2 Derives Meaning			Criterion 3 Solves Problem			Criterion 4 Looks Back		
	1	2	3	1	2	3	1	2	3	1	2	3
1 4 Steps To Solve It	•	•	•	•	•	•	•	•	•	•	•	•
2 Step by Step	•	•	•	•	•	•	•	•	•	•	•	•
3 Graphic Organizers	•	•	•	•	•	•	•	•	•	•	•	•
4 Log and Checklist	•	•	•	•	•	•	•	•	•	•	•	•
5 Mnemonics	•	•	•	•	•	•	•	•	•	•	•	•
6 Everyday Math	•											
7 Write It, Explain It	•											
8 Get Real	•											
9 Define the Word or Symbol		•										
10 Vocabulary Games		•										
11 Word Clues		•										
12 Math Jeopardy		•	•									
13 Highlighting			•									
14 What's Missing?			•									
15 What's the Question?			•									
16 Can't Tell			•									
17 Get Graphic				•								
18 Use Manipulatives				•	•	•						
19 Draw a Diagram				•	•	•						
20 Find a Pattern				•	•	•		•			•	•
21 Make a Table				•	•	•		•			•	•
22 Make a List				•	•	•		•			•	•
23 Work Backward				•	•	•		•			•	•
24 Guess and Test				•	•			•			•	•
25 Estimation Games		•	•				•					
26 The Price Is Right							•					
27 Beginning Problem Solving					•			•	•			•
28 Which Operation?								•				
29 Formulas								•				
30 Maze It or Cloze It								•				
31 Peer Partners								•			•	•
32 Assigning Students to Groups	•	•		•							•	•
33 Collaborative Problem Solving	•	•	•	•	•	•	•	•	•	•	•	•
34 Use a Matrix	•	•	•	•	•	•	•	•	•	•	•	•

3

Figure 1. Continued.

Intervention Plan

BEHAVIOR	INTERVENTION IDEA NUMBER	START DATE	NOTES
❶ Formulate steps to solve problems	1 2	Oct 10 Oct. 14	Use both of these as whole-class activities.
❷ Use specific strategies to solve problems	19	Oct. 25	Set up center with specific strategies. Begin with Draw a Diagram.
❸ Continue with Ideas 20-24			
❹			
❺			
❻			

Additional Assessment Information

Record any other assessment information that exists on the student related to math problem-solving performance.

Standardized Assessment

Comprehensive Mathematical Abilities Test
(CMAT) 91

Informal Assessment
(observations, interviews, curriculum-based assessment, rating scales, checklists, student work samples)

Doesn't know how to begin solving problems.
Seldom uses strategies to solve problems.

Current GPA in math is 32.

Figure 1. Continued.

Idea 1
4 Steps To Solve It

Solving math problems involves a number of steps. George Polya developed a four-step problem-solving model that is one of the most well-known math problem-solving models in the world. Here are his four steps:

① understand the problem
② devise a plan
③ carry out the plan
④ look back

Post or project the 4 Steps To Solve It so that all students can see it, and use it to work through several problems so that students understand how to apply the steps.

We have provided two sets of self-reflective questions that guide each step. The first set is for younger students, and the second set is for older students. Laminate the small cards for students to keep in their math folders.

> **Tip:**
> Use sample problems completed by students (make sure to delete the student's name) to demonstrate how to apply each step. Choose a variety of problems demonstrating various levels of quality. For example, ask students if the clue words in the example problem are underlined. If they are not, underline them. Continually link the four-step process to solving the problem.

Note. This idea is based on the model from *How To Solve It: A New Aspect of Mathematical Method* (2nd ed.), by G. Polya, 1957, Princeton, NJ: Princeton University Press. Copyright 1957 by Princeton University Press.

4 Steps To Solve It

① **Understand the Problem**

Did I draw a line under clue words?

What numbers are in the problem?

What is the problem asking?

② **Devise a Plan**

Have I ever done another problem like this before?
Can I draw a picture or make a table?
Did I estimate the answer?

③ **Carry Out the Plan**

Have I used all the important numbers from the problem?
Did I show my work?
Can I explain my answer?

④ **Look Back**

Does my answer make sense?
Did I answer the question?
Did I label my answer?

4 Steps To Solve It

① Understand the Problem

Did I read the entire problem carefully?

Have I underlined clue words?

What facts and data am I given?

What question do I need to answer?

② Devise a Plan

Can I restate the problem in my own words?

Have I seen a problem like this before?

Can I draw a picture or make a table to help me solve the problem?

③ Carry Out the Plan

Does this plan seem reasonable?

Have I used all the data to obtain the answer?

Have I checked each step of my plan?

Will I be able to prove how I arrived at my answer?

④ Look Back

Does my answer make sense?

Did I answer the question?

Did I answer using the language in the question?

Did I place the correct label on my answer?

4 Steps To Solve It

① **Understand the Problem**

② **Devise a Plan**

③ **Carry Out the Plan**

④ **Look Back**

---

4 Steps To Solve It

① **Understand the Problem**

② **Devise a Plan**

③ **Carry Out the Plan**

④ **Look Back**

---

4 Steps To Solve It

① **Understand the Problem**

② **Devise a Plan**

③ **Carry Out the Plan**

④ **Look Back**

---

4 Steps To Solve It

① **Understand the Problem**

② **Devise a Plan**

③ **Carry Out the Plan**

④ **Look Back**

© 2006 by PRO-ED, Inc.

Idea 1

Idea 2
Step by Step

One of the most frequently used techniques to help students solve math problems is a form that directs students through the steps that are important to arrive at a correct answer. In Idea 1, 4 Steps To Solve It, we provide a four-step problem solving model. This idea is an elaboration of Idea 1 and provides a form that can be used as further guidance.

Here is how it works.

1. Give students a copy of the Step by Step form.
2. Model for the students how to complete the form.

- Determine what you need to find out.
- Write down key words.
- State what you know and what you do not know.
- Draw a picture or make a table or diagram.
- Circle the operation you need to perform.
- Solve the problem.
- Check *yes* if the answer obtained makes sense and *no* if it does not.
- Write the answer and make sure to include a label.

Step by Step

- **What do I need to find?**
  The perimeter of the garden

- **Key Words**
  triangle   three
  perimeter

- **What I know**
  The perimeter is the distance around a figure.   The length of each side.
  A triangle has three sides.

- **What I do not know**
  The measurement of the perimeter of the triangle.

- **Draw It** (picture, table, or diagram)

  23.64 ft   7.41 ft
      26.3 ft

- **Operation:**   (+)   −   ×   ÷

- **Solve:**
  ```
    23.64
    26.30
  +  7.41
  -------
    57.35
  ```

- **Does my answer make sense?** ☑ Yes ☐ No

- **Answer with label:** 57.35 feet

> Katy wants to put a fence around her garden, which is in the shape of a triangle. The three sides are 23.64 ft, 7.41 ft, and 26.3 ft. How many feet of fencing does she need to purchase?

Step by Step

✪ What do I need to find?

✪ Key Words

✪ What I know

✪ What I do not know

✪ Draw It (picture, table, or diagram)

✪ Operation: +   −   ×   ÷

✪ Solve:

✪ Does my answer make sense?   ✪ Answer with label:
　❏ Yes   ❏ No

© 2006 by PRO-ED, Inc.   Idea 2

Idea 3
Graphic Organizers

Using graphic organizers and other visual strategies is always a good idea when teaching students to become better math problem solvers. This idea presents a number of graphic organizers and tools that will help students organize information and demonstrate their understanding of the math problem-solving process. For each graphic organizer we have provided directions and a brief explanation.

Group Talk

Group Talk is a graphic organizer that helps students learn the problem solving process in a nonthreatening atmosphere. It can also be used as a brainstorming activity for students once they become comfortable with the process.

Here's how it works.

1. Present a problem to the class by projecting it or writing it on the board.

2. Ask students to determine what the problem is asking and to write the question in the middle rectangle on the Group Talk graphic organizer, under "What question are you trying to answer?"

3. Complete the rest of the organizer, starting with rectangle #1 and proceeding clockwise.

Tip:
Use group talk as a brainstorming activity. Ask students to read a problem or a passage from the text. Have them tell you the main idea and write it in the middle rectangle on the blank Group Talk graphic organizer. Ask students to brainstorm to fill in the other rectangles with additional information, such as definitions, real-world examples, operation needed, and important details.

Note. Take These Steps is from *Practical Ideas That Really Work for Students with ADHD: Grade 5 Through Grade 12* (2nd ed., p. 131), by K. McConnell and G. R. Ryser, 2005, Austin, TX: PRO-ED. Copyright 2005 by PRO-ED, Inc. Reprinted with permission. Chart It is from *Practical Ideas That Really Work for Students with Dyslexia and Other Reading Disorders* (p. 136), by J. Higgins, K. McConnell, J. R. Patton, and G. R. Ryser, 2003, Austin, TX: PRO-ED. Copyright 2003 by PRO-ED, Inc. Reprinted with permission.

Take These Steps

Take These Steps is a simple form to help students learn the steps in a process. Use it to write down the steps in a problem-solving process or in a procedure.

Here's how it works.

❶ Present a process or procedure to the class (e.g., steps in a problem-solving strategy or steps on how to cross-multiply).

❷ Ask students what the first step in the process or procedure is and tell them to write it or draw a picture illustrating it in the first box.

❸ Complete the rest of the steps with the class.

❹ Allow students to use the form as a study guide while completing homework or studying for a test.

Chart It

Using charts is a wonderful way to help students organize and make sense of information found in math problems. The chart provides students with a focus and helps them know what information is important and needs to be remembered.

Here's how it works.

❶ Use the chart when working on one type of problem (e.g., double-digit addition, complementary angles).

❷ Write the first problem in the column under the Problem heading.

❸ Ask the students to state the important information and list it in the column under the Important Information heading.

❹ Complete the rest of the chart through discussion and modeling.

Labels, Labels, Everywhere

Many students who are poor math problem solvers seldom think about relating their final answer to the original problem. Providing boxes for them to place the label reminds students to take that final step, translating the final number into the context of the original problem. The forms provided include a box for the final answer and a box for the appropriate label. The first form should be used for money problems, the second form for other types of problems.

Examples

Problem

Six people can pick 50 quarts of strawberries in 1 day. A farmer gets an order for 175 quarts of strawberries for the next day.
How many people should she hire?

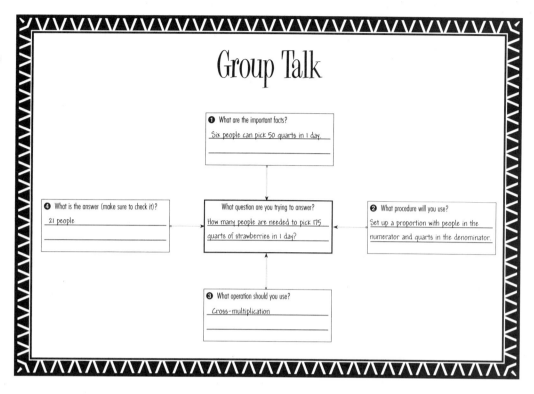

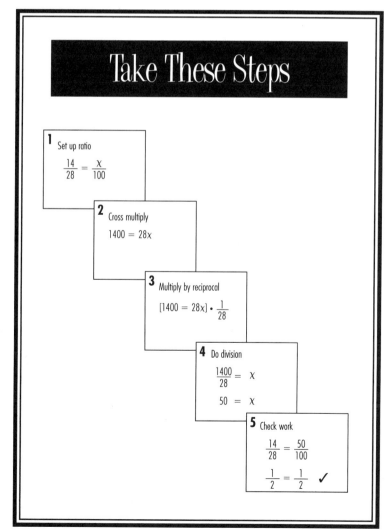

Examples

Chart It

Complementary Angles
Type of Problem

Problem	Important Information	Operation	Solution
$m\angle 1 = 2x$ $m\angle 2 = 3x$	$\angle 1$ and $\angle 2$ are complementary angles. $m\angle 1 + m\angle 2 = 90°$	$2x + 3x = 90$ $5x = 90$ $\dfrac{5x}{5} = \dfrac{90}{5}$	$x = 18$
Two angles are complementary. The measure of one is 5 times the measure of the other. Find the measure of each angle.	Complementary angles sum to 90°. $m\angle 1 = 5m\angle 2$ $m\angle 1 = x$ $m\angle 2 = 5x$	$x + 5x = 90$ $6x = 90$ $\dfrac{6x}{6} = \dfrac{90}{6}$	$x = 15$

Labels, Labels, Everywhere

Problem Number 5

Jamal is saving for a new CD. He has half of what he needs. He has $10.25. How much does the CD cost?

$$\begin{array}{r} 10.25 \\ \times\ 2 \\ \hline 20.50 \end{array}$$

Label: $ | Answer: 20.50

Problem Number 7

This month Mark saved 5 times as much money as he did last month. If he saved $35.00 this month, how much did he save last month?

$$5\overline{)35}\ \ ^{7}$$

Label: $ | Answer: 7.00

Labels, Labels, Everywhere

Page 86 **Problem Number** 15

Seth bought 4 pairs of pants and 3 shirts. How many different outfits can he make?

$4 \times 3 = 12$

Answer: 12 | Label: outfits

Page 86 **Problem Number** 16

The ferry can hold 8 cars. How many trips will it have to make to carry 27 cars across the river?

$$8\overline{)27}\ \ ^{3\frac{3}{8}}$$
$$\dfrac{24}{3}$$

Answer: 4 | Label: trips

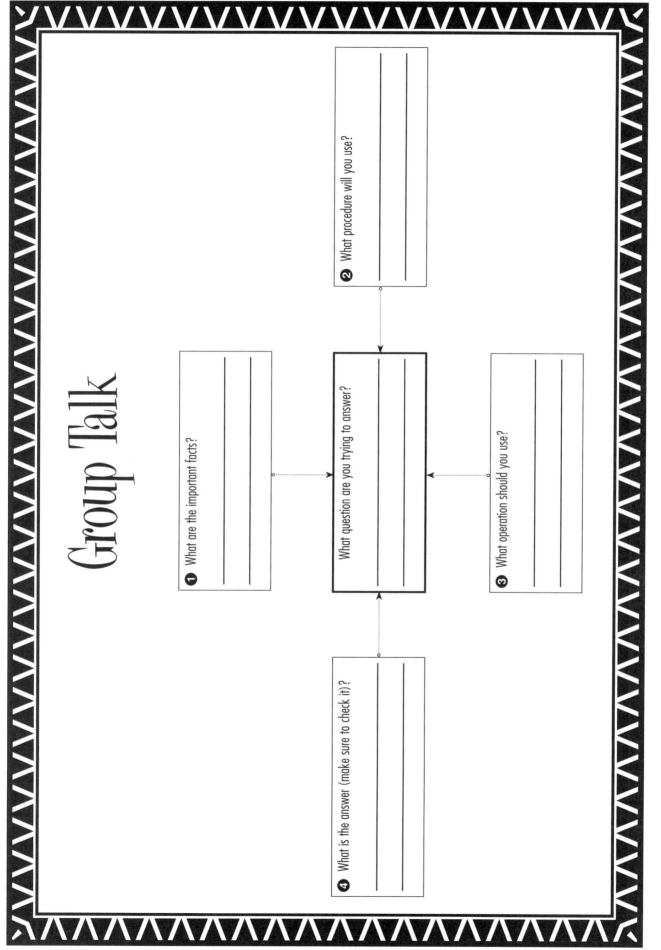

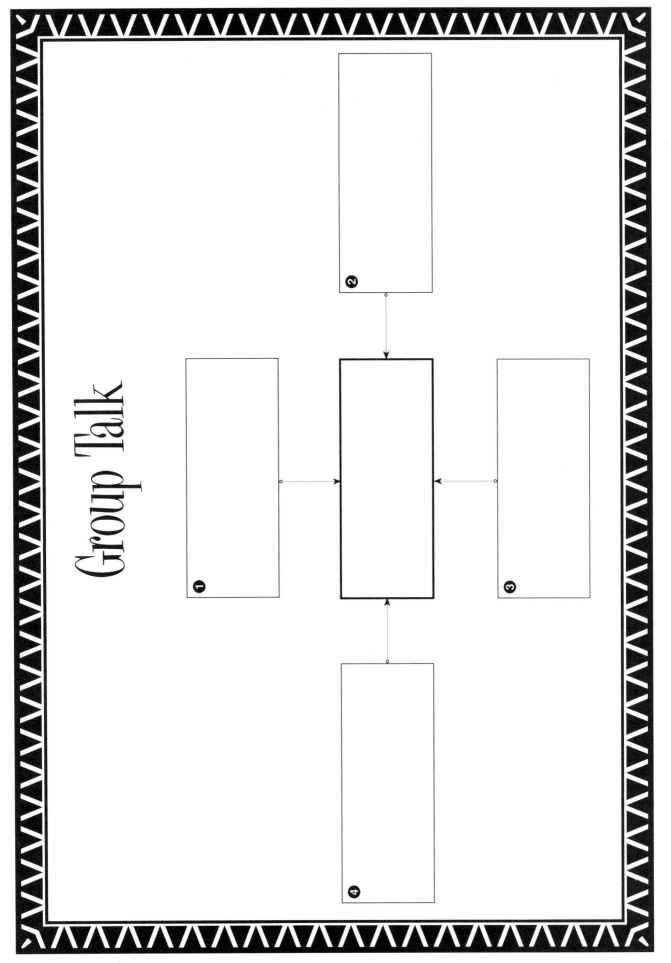

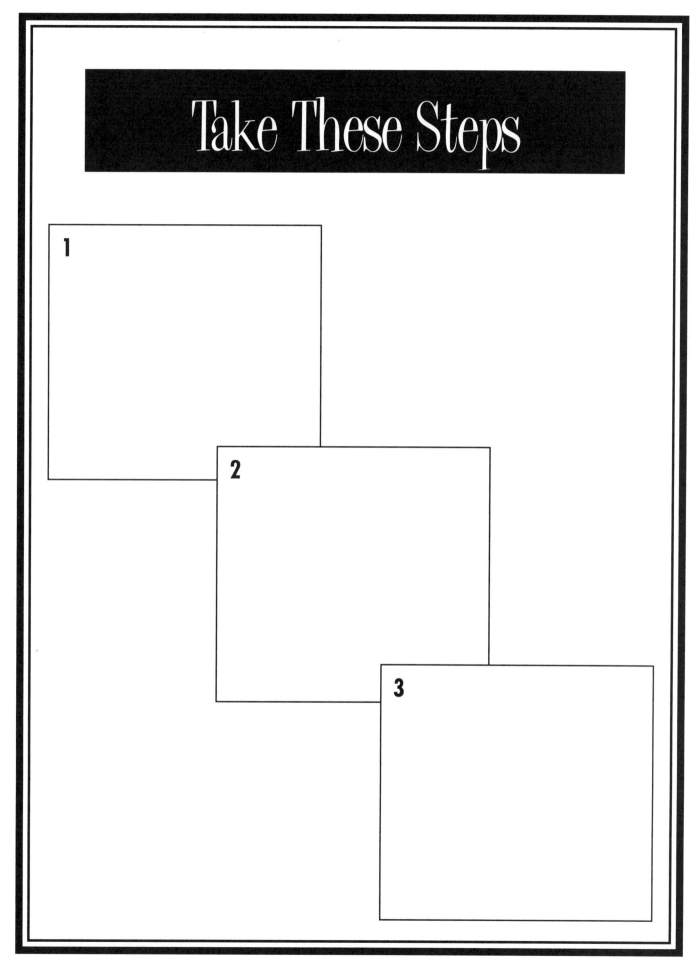

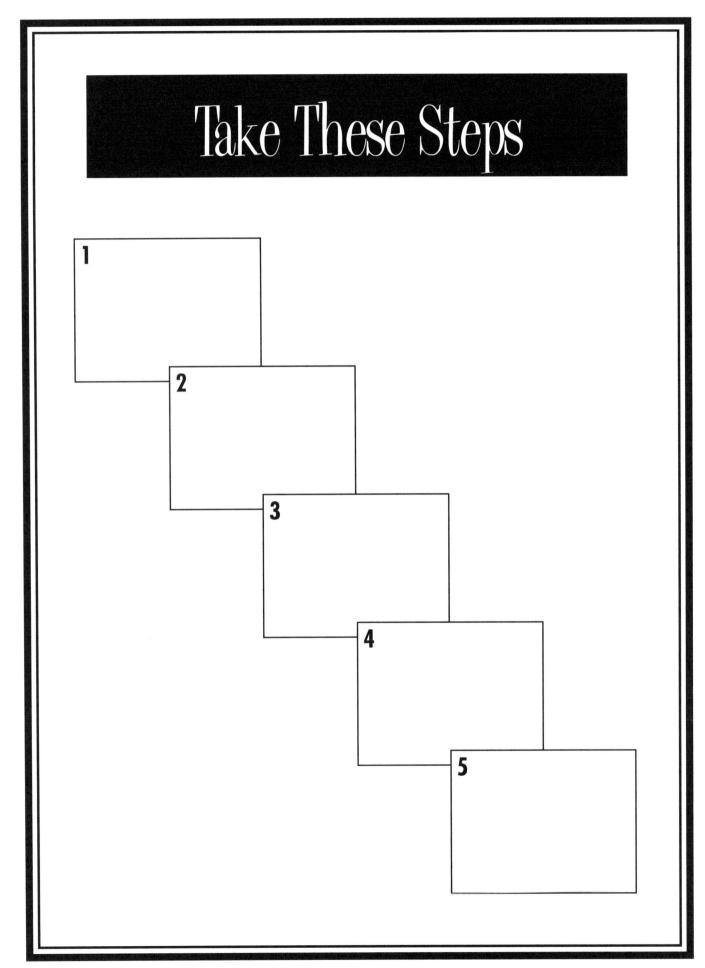

Chart It

Type of Problem _____

Problem	Important Information	Operation	Solution

Labels, Labels, Everywhere

Problem Number _____

$ Label   Answer

Problem Number _____

$ Label   Answer

Problem Number _____

$ Label   Answer

Problem Number _____

$ Label   Answer

© 2006 by PRO-ED, Inc.

Idea 3

Labels, Labels, Everywhere

Page _____	Problem Number _____

$ Label | Answer

Page _____	Problem Number _____

$ Label | Answer

Page _____	Problem Number _____

$ Label | Answer

Page _____	Problem Number _____

$ Label | Answer

Page _____	Problem Number _____

$ Label | Answer

Page _____	Problem Number _____

$ Label | Answer

© 2006 by PRO-ED, Inc.

Idea 3

Labels, Labels, Everywhere

Problem Number _____

Answer | Label

Problem Number _____

Answer | Label

Problem Number _____

Answer | Label

Problem Number _____

Answer | Label

© 2006 by PRO-ED, Inc.

Idea 3

Labels, Labels, Everywhere

Page ______ Problem Number ______ Answer: [ ] Label: [ ]	Page ______ Problem Number ______ Answer: [ ] Label: [ ]
Page ______ Problem Number ______ Answer: [ ] Label: [ ]	Page ______ Problem Number ______ Answer: [ ] Label: [ ]
Page ______ Problem Number ______ Answer: [ ] Label: [ ]	Page ______ Problem Number ______ Answer: [ ] Label: [ ]

© 2006 by PRO-ED, Inc.

Idea 4
Log and Checklist

To be efficient math problem solvers, students need to follow some clear procedures. Use the Problem Solving Log to determine if students are skipping steps, then use the Math Checklist, provided, or develop a personalized math checklist.

> ### Try these methods:
>
> ❶ Use the Problem Solving Log to record the steps a particular student completes for five types of problems. With the student, review the log and use the blank math checklist to write questions that reflect the steps the student typically skips. For example, if the student did not check his or her work for at least three of the five problems, one question to include is, "Did I check my answer to make sure it is correct?"
>
> ❷ Give each student in the class a blank Problem Solving Log. For five different assignments, choose one problem and have the students place an X next to each step they complete for that problem. When all five problems have been completed, set up individual teacher talk times to review the log.
>
> ❸ Have each student complete a Problem Solving Log. Using the Math Checklist, provided in this idea, review the questions with the entire class, and model using it for several types of problems. Next, have all students use the checklist to complete at least five problems. Upon completion, have the students complete a second log, and compare it to their first log. Check each student's log to determine if there was improvement.

Tip:
For best results, complete the log using a variety of problems across a 1- to 2-week period.

Problem Solving Log

Name: _____ Date: _____

Place an X next to each step under the correct problem if the step was completed.

Step	Problem ____	Problem ____	Problem ____	Problem ____	Problem ____
Wrote down the important information in the problem.					
Used a diagram, table, or picture to help solve the problem.					
Estimated the answer.					
Used the correct operation to solve the problem.					
Used the correct label on the answer.					
Checked answer. If incorrect, redid problem.					
Used what was done in this problem to help solve a different problem.					

© 2006 by PRO-ED, Inc.

Idea 4

Problem Solving Log

Name: _____ Date: _____

Place an X next to each step under the correct problem if the step was completed.

Step	Problem ___	Problem ___	Problem ___	Problem ___	Problem ___

© 2006 by PRO-ED, Inc.   Idea 4

Math Checklist

Name: _____   Date: _____

Check off each question as you complete it.

❏ Did I write down important information from the problem so that I could become familiar with it?

❏ Did I write down the numbers correctly?

❏ Did I use a diagram, picture, or table to help me complete the problem?

❏ Did I use the correct operation or procedure to answer the problem?

❏ Were my answer and estimate close?

❏ Did I check my answer to make sure it was correct?

❏ Did I use the correct label on my answer?

❏ _____?

❏ _____?

❏ _____?

Math Checklist

Name: _____ Date: _____

Check off each question as you complete it.

☐ _____ ?

☐ _____ ?

☐ _____ ?

☐ _____ ?

☐ _____ ?

☐ _____ ?

☐ _____ ?

☐ _____ ?

☐ _____ ?

☐ _____ ?

© 2006 by PRO-ED, Inc.   Idea 4

Idea 5
Mnemonics

Mnemonics are memory devices that are helpful to all students, particularly those who have difficulty learning content. This idea contains three mnemonic acronyms that can help students remember the steps they should take when trying to solve math problems. To be most effective, make sure that students become automatic in their recall of the acronym.

Here are the steps for teaching students to use a mnemonic.

1. Present one of the mnemonics and demonstrate its usefulness by modeling the individual steps and explaining their use and relevance.

2. Have students learn the mnemonic through recitation and repetition.

3. Provide problems that are mathematically easy for students to practice using the mnemonic.

4. Provide grade-level problems for students to practice using the mnemonic with more challenging math problems.

5. Review the mnemonic periodically to ensure it has been maintained.

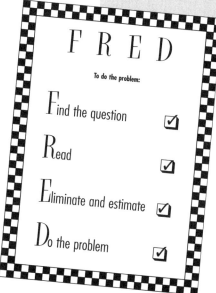

Ways To Use Mnemonics

- Choose one mnemonic and post it. Leave it posted until students are using it independently, then post a different mnemonic.

- Let students choose the mnemonic that they think is most helpful. Laminate the mnemonic card for them to keep in their math folders.

- Require students to do at least part of their homework using one of the mnemonic checkoff sheets.

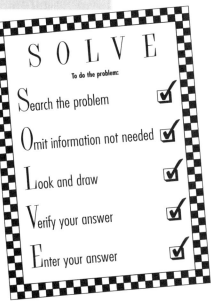

Note. RED and FRED are from *Practical Ideas That Really Work for Students with ADHD: Grade 5 Through Grade 12* (2nd ed., p. 111), by K. McConnell and G. R. Ryser, 2005, Austin, TX: PRO-ED. Copyright 2005 by PRO-ED, Inc. Reprinted with permission. SOLVE is adapted from "Mnemonics: Enhancing the Math Performance of Students with Learning Difficulties," by S. P. Miller and C. D. Mercer, 1993, *Intervention in School and Clinic, 29,* pp. 78–82. Copyright 1993 by PRO-ED, Inc. Adapted with permission.

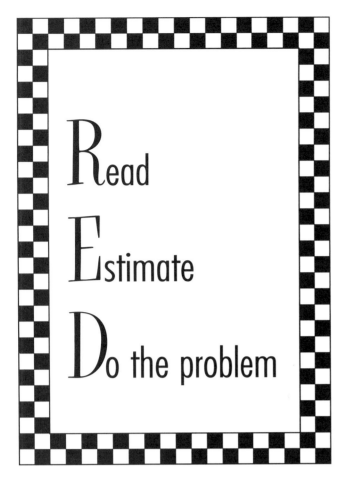

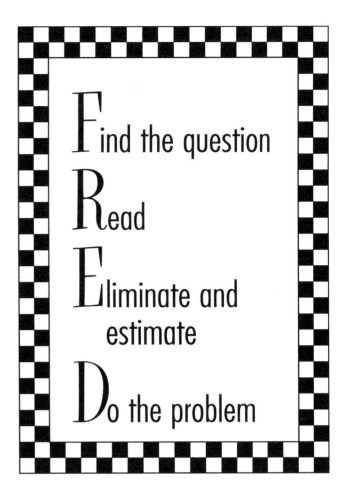

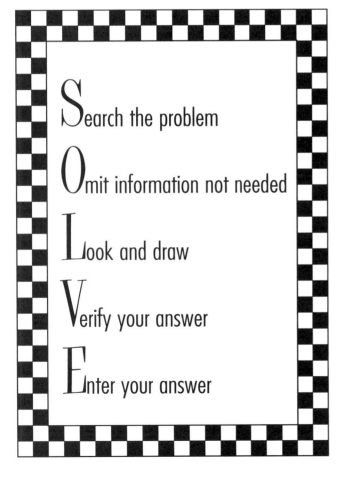

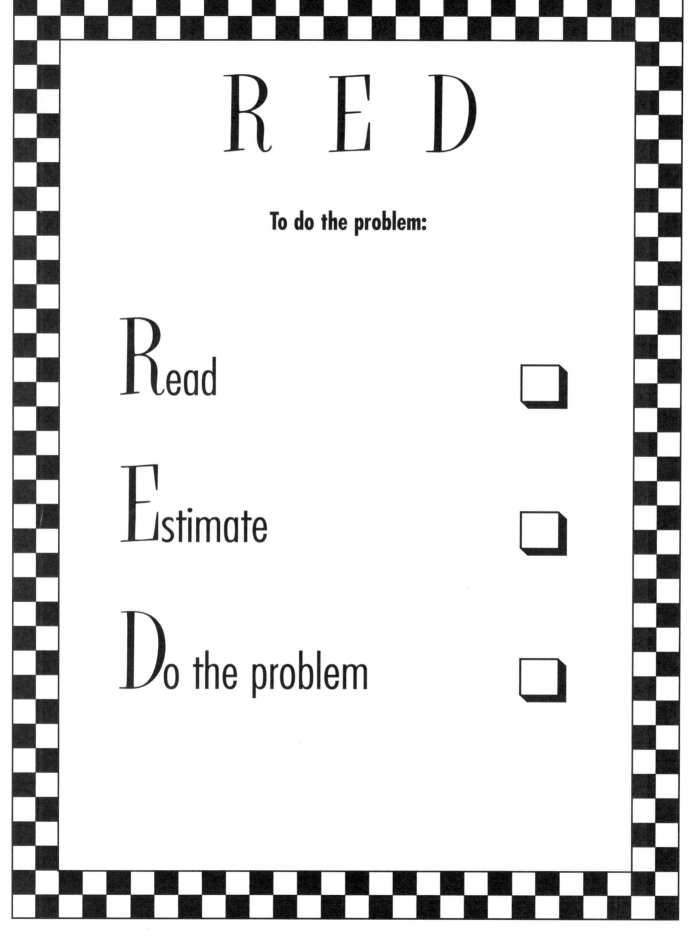

FRED

To do the problem:

Find the question ☐

Read ☐

Eliminate and estimate ☐

Do the problem ☐

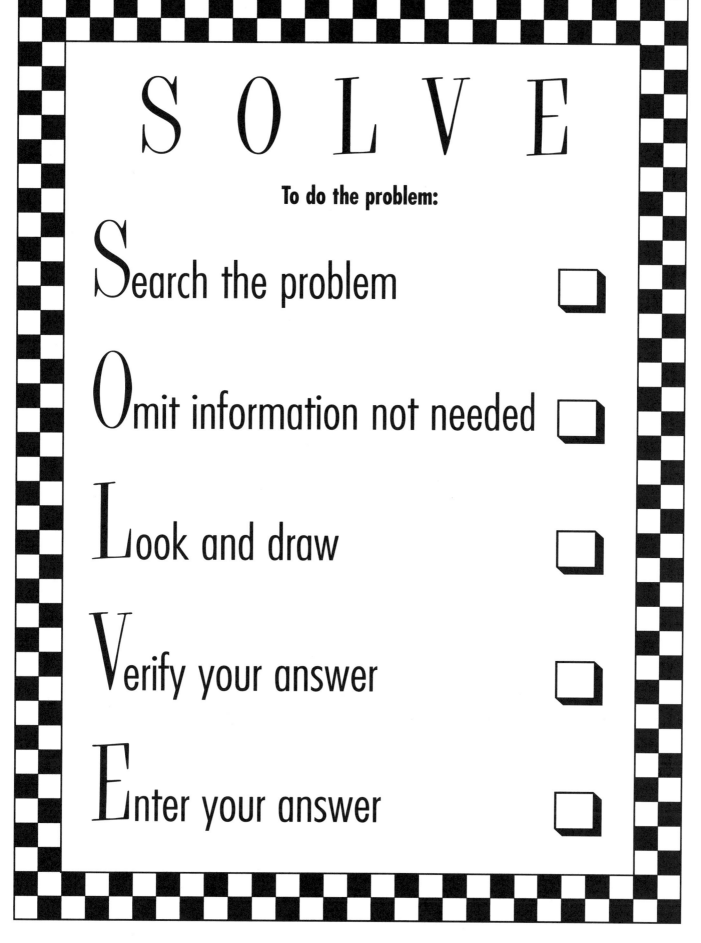

Idea 6
Everyday Math

Math is all around us. We use math, in some fashion, from the time we get out of bed until we go to sleep at night. Most of the math we use is not algebra, geometry, trigonometry, or calculus—it is everyday, practical math that typically involves some form of measurement of time, money, temperature, volume, weight, or length. It is important to point out to students how relevant math is in their everyday lives and that math problem solving takes many different forms in everyday life. This activity provides a simple way to engage students in the discussion of how they use math on a daily basis.

Here's how to get students to think about everyday math.

❶ Tell students that you want them to think about every way they use math in their lives—*outside of math class* and *math homework assignments*.

❷ Give them a copy of the form A Math Day in the Life Of to complete as an activity (e.g., as a homework assignment).

❸ Create a master chart for the entire class on which you can enter student results.

❹ Discuss the math demands that are identified by the students.

Tip:
Group the results using categories such as cooking, sports, time, and so forth. Once the results have been grouped, find the percentage for each group. Graph the results on graph paper.

A Math Day in the Life Of

_____

List anything that you do that involves math for the following times of day.

Early Morning (before school)

- _____
- _____
- _____

At School (not in math class)

- _____
- _____
- _____

After School (in the afternoon when school is over)

- _____
- _____
- _____

Evening or Night (does not include math homework)

- _____
- _____
- _____

Question: What is the one type of math skill that you use the most?

- _____

© 2006 by PRO-ED, Inc.

Idea 6

Idea 7
Write It, Explain It

In Idea 6, Everyday Math, we presented an activity to engage students in thinking about math in everyday life. Write It, Explain It goes a step further and provides a method for students to develop problems from the Everyday Math activity. An added bonus is that the teacher can gauge students' depth of understanding by asking them to present the solution to their problem.

Here's how it works.

❶ Copy the cards, front and back. After completing Everyday Math, each student writes a problem, using their personal experiences, on the front of the card with the solution on the back, and turns it in to the teacher.

❷ Read through the cards to make sure that the answer and the procedure used to arrive at the answer are correct. Meet privately with students who have an incorrect answer or procedure to work through it.

❸ Choose one of the cards at random, place it on a document viewer, and ask students to answer the problem.

❹ When everyone is finished, the author of the problem provides the answer and explains how he or she reached the solution.

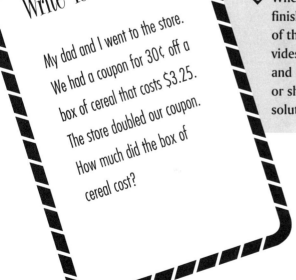

Write Your Problem

My dad and I went to the store. We had a coupon for 30¢ off a box of cereal that costs $3.25. The store doubled our coupon. How much did the box of cereal cost?

Show Your Solution

```
  30¢
×  2
  60¢
```
This is the amount we saved because the coupon was doubled.

```
 $3.25
− 0.60
 $2.65
```
Our cereal costs $2.65.

Idea 7

Show Your Solution

Show Your Solution

Show Your Solution

Show Your Solution

© 2006 by PRO-ED, Inc.

Idea 7

Idea 8
Get Real

Often students do not understand the value of mathematics because they cannot relate it to real life. This idea requires students to identify the question in a math problem from the text and then write a similar problem from real life.

Here's how it works.

❶ Give each student a copy of the form Get Real.

❷ Project a problem from the text or unit onto a whiteboard or other appropriate space.

❸ Identify the question in the problem, and write it in the left column.

❹ Ask students to write a similar, real-life problem in the right column.

❺ Once students are comfortable with the process, allow them to choose three or four problems from each unit of study on which to complete the Get Real form.

Tip:
If students have difficulty writing, allow them to draw a picture of the real-life problem and write in the numbers needed.

Get Real

Identify the question in the problem.	Write a similar problem from real life.
Page _____ Problem _____	
Page _____ Problem _____	
Page _____ Problem _____	
Page _____ Problem _____	

Idea 8

Idea 9
Define the Word or Symbol

Understanding and using correct mathematical vocabulary and symbols is a critical skill for good math problem solving. To help your students learn new math vocabulary and symbols, we suggest two practical strategies that can be used on an ongoing basis:

- word walls
- student word books

Both of these strategies are simple to implement and require little in the way of materials. You can even use both at the same time.

To use Define the Word or Symbol, follow these directions.

Word Walls

1. To create a word wall, clear a large space on a blank wall.

2. Write the letters in alphabetical order on a chart or on cards that are attached directly to the wall. Include an additional card for symbols.

3. As words or symbols are introduced or used, place them (also on cards) under the letter that is the same as the first letter of the word or under the symbol card.

4. Continue to add to the word wall lists throughout the year.

Student Word Books

1. Make a copy of the Define the Word or Symbol worksheet for students to keep in their math folders.

2. Complete the worksheet as a group the first several times by identifying new math vocabulary or symbols in problems or from the text.

3. Write the new words or symbols in column one; write the text or dictionary definition of the word or symbol in column two; rewrite the definition using the students' own words in column three.

4. Complete the worksheets independently once students understand the directions.

> ### 🌀 Tips:
> - Index cards are an alternative format to books and pages. Students can write each new word they learn on a card. On the back of the card, they can write an equivalent word in their language, and draw an illustration of the word. By punching a hole in the cards and gathering them on a key ring, students can create a set of flash cards that functions as a word book.
> - Allow students to use their worksheets or index cards during tests as a reward for their completion.

Define the Word or Symbol

Word or Symbol	Definition	Definition in Your Own Words
rational number	A number that can be put in the form $\frac{a}{b}$, where a and b are integers and $b \neq 0$.	Any two integers that can be divided (divisor $\neq 0$): $\frac{1}{2}, \frac{-8}{5}, \frac{3}{1}$
reciprocal	The rational number that multiplies r (r ≠ 0) to give $\frac{1}{r} \cdot r = 1$.	Two rational numbers that equal 1 when multiplied: $\frac{2}{3} \cdot \frac{3}{2} = \frac{6}{6} = 1$
ratio	Quotient of two numbers: $\frac{a}{b}$	Two numbers that are divided: $\frac{3}{4}$
proportion	Equality between two ratios: $\frac{a}{b} = \frac{c}{d}$	Two fractions that are equal: $\frac{1}{2} = \frac{5}{10}$

Define the Word or Symbol

Word or Symbol	Definition	Definition in Your Own Words

Front of Cards

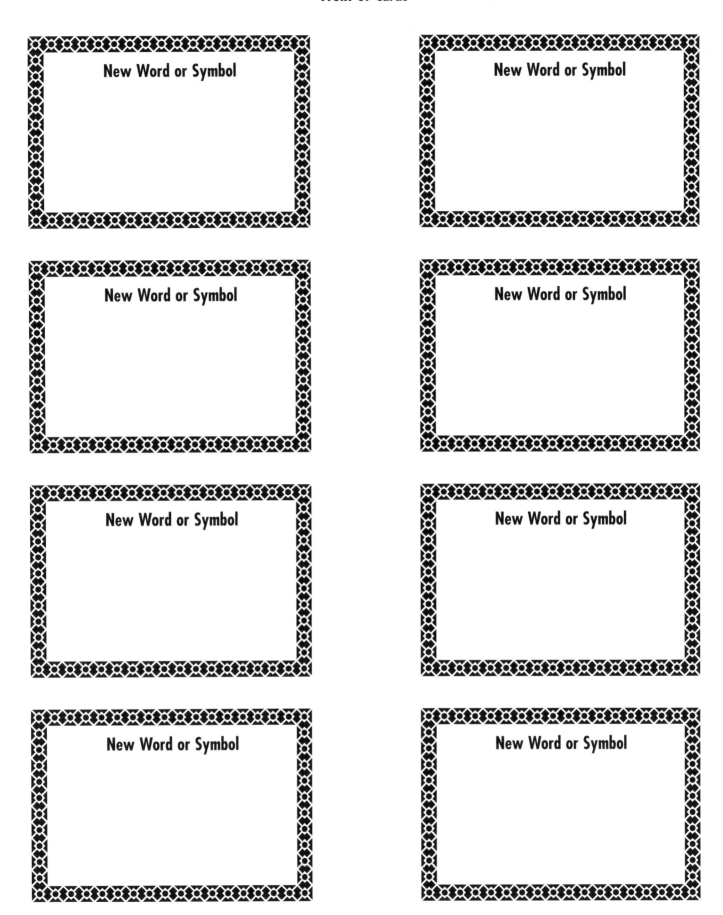

© 2006 by PRO-ED, Inc. Idea 9

Back of Cards

Idea 10
Vocabulary Games

Understanding the vocabulary that is part of the math curriculum is problematic for many students. Until students are comfortable using new math vocabulary words, they will need to practice in structured situations. An effective way to teach math vocabulary is through the use of vocabulary card-deck games. Here are three simple vocabulary games that are familiar to most students. All of these games provide a great way to practice words so that students understand their meanings.

Game 1: Fishing for Words
This game works best with at least 10 vocabulary words.

❶ Before beginning the lesson, write the vocabulary words on the card templates provided and make four copies of each math vocabulary word. Put all the cards together into a deck and shuffle.

❷ Divide the students in to groups of four, and choose a dealer who will deal five cards to each student in the group. Scatter the remaining cards face down in the center of the table (or floor or wherever the game is being played).

❸ The student to the dealer's left begins by picking one word from his or her cards, saying it aloud, and giving a brief definition of it. If the student defines the word appropriately, he or she can then ask any student in the group to "Give me all of (the specific word)." If the student does not define the word correctly, he or she loses the turn.

❹ Students who are asked must give up their word cards, if they have them; if they do not have the word card, they direct the student who asked to "Go fish" in the center pile.

❺ The student can keep asking other students for cards until they are unable to define the word, or the person who they ask for a card tells them to "Go fish."

❻ The sequence is repeated until someone has no cards left. The student with the most sets (of four cards that have the same word) wins.

Tip:
These games can be modified to focus just on word recognition, if needed.

Note. This idea is adapted from *Practical Ideas That Really Work for Students with Dyslexia and Other Reading Disorders,* by J. Higgins, K. McConnell, J. R. Patton, and G. R. Ryser, 2003, Austin, TX: PRO-ED. Copyright 2003 by PRO-ED, Inc. Adapted with permission.

Game 2: Word Match
This game can be played with the entire class or in small groups.

❶ Put the students in small groups and give each group the same set of 6 to 10 vocabulary cards.

❷ The leader (the teacher or a student) gives a clue to the meaning of the word by saying the word or phrase that is associated with one of the vocabulary words. (Note: A sheet can be prepared ahead of time with the vocabulary words, a key phrase, and the complete definition.)

❸ The first group to associate the clue with the correct vocabulary word turns in their card with the word.

❹ The first group to run out of cards wins.

Game 3: Word Concentration
This game is designed for pairs of students.

❶ Develop two card decks: One deck contains the vocabulary word and the other deck has the definitions. (Note: The backside of each deck should be blank or have some type of design.) We have provided a sample of both types of decks and a designed set to copy for the backs.

❷ Place the cards on a table or on the floor in a quiet area.

❸ Two students place the cards (the vocabulary and its matching definition) face down and scramble the cards.

❹ Player 1 turns over two cards. If the cards match (a match is a vocabulary word and its corresponding definition) and Player 1 identifies it as a match, he or she takes the two cards and may take another turn. If the cards do not match, it becomes Player 2's turn.

❺ Play continues until all cards are matched.

❻ At the end of the game, the player with the most pairs of cards wins.

Word Cards

decimal number	numerator
factor	denominator
integer	ratio
fraction	reciprocal

© 2006 by PRO-ED, Inc.

Idea 10

Definition Cards

a number written to base 10	In the fraction $\frac{x}{y}$, x is the _____.
an exact divisor of a number	In the fraction $\frac{x}{y}$, y is the _____.
positive and negative whole numbers and zero	the quotient of two numbers
an expression of the form $\frac{x}{y}$	For a number x, the _____ is $\frac{x}{y}$

Blank Cards

Backs for Cards

© 2006 by PRO-ED, Inc.

Idea 10

Idea 11
Word Clues

Key words in math problems provide clues to students about how to solve them. Many students can perform an algorithm when it is presented in isolation (e.g., 2 + 2), but become confused when presented with a word problem that must be solved using that same algorithm. This idea presents a method to teach students to solve problems by looking for key words. These words often indicate the operation they should use to solve the problem. Cards with key words are provided for addition, subtraction, multiplication, and division. There is also a page of blank cards so that you and your students can add words to each set.

Here's how it works.

❶ Copy and laminate one card for each student that matches the level you are teaching.

❷ Give each student the first card.

❸ Present two or three word problems that use the first key word and solve them as a group.

❹ Have students solve additional problems, highlighting the key words each time.

Try these variations.

Game
Divide the class into groups of three to four students. Give each group the four operation cards (addition, subtraction, multiplication, and division) that are provided at the end of this idea. Project a word problem containing a word clue. Have the group raise the card that shows the operation that should be used to solve the problem. If desired, 1 point can be awarded to the group that raises the correct card first.

Note. Sometimes key words do not lead to the operation under which they are listed in this idea. As students become more proficient math problem solvers, move away from this strategy.

Center

Copy and laminate one set clue words. Include several word problems that have the clue words highlighted and several more for the student to highlight. Have the student make up one or more problems using the clue words (See Idea 7—Write It, Explain It—for problem cards).

Chart

Post the Operation Word Clues chart or provide a copy to each student. Encourage students to add new word clues under the appropriate headings. As reinforcement, allow students to use their charts during tests.

Addition

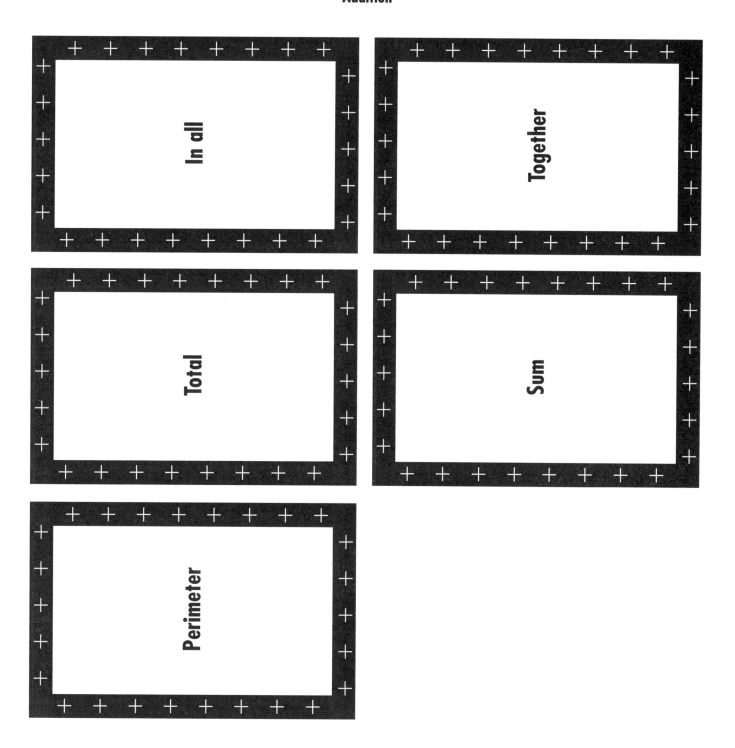

Subtraction

- Left
- How many more
- How much more
- Fewer
- Difference
- Begin with
- Took away
- Exceed

Multiplication

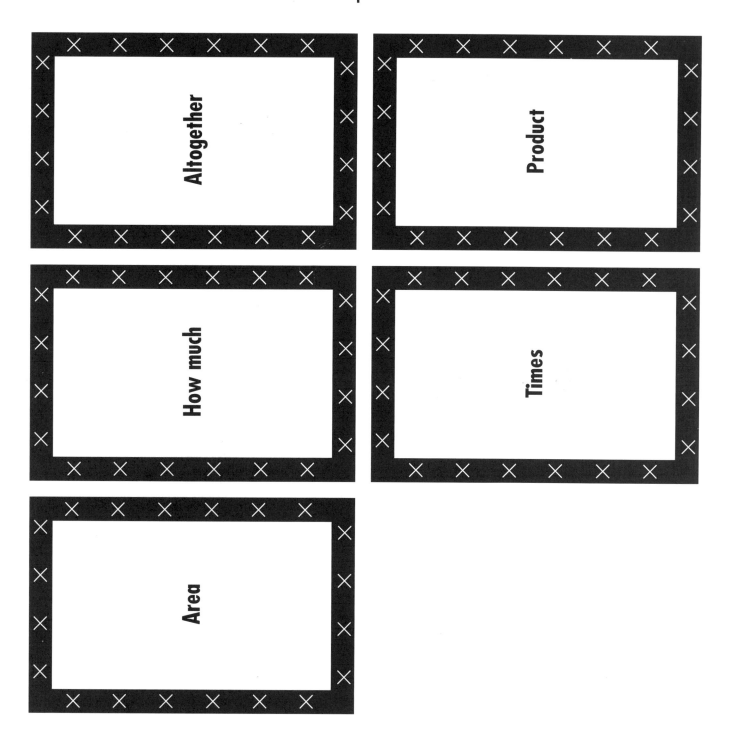

© 2006 by PRO-ED, Inc.   Idea 11

Division

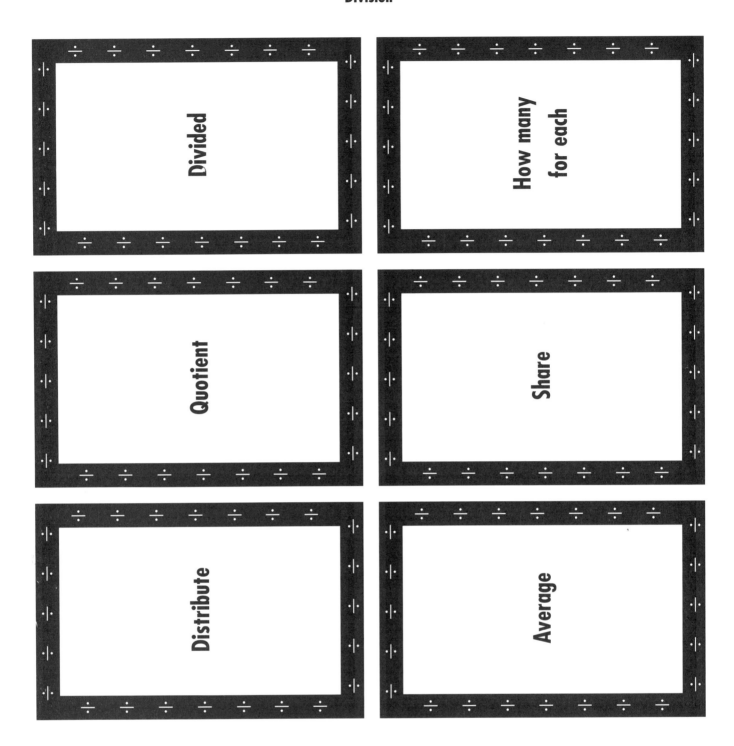

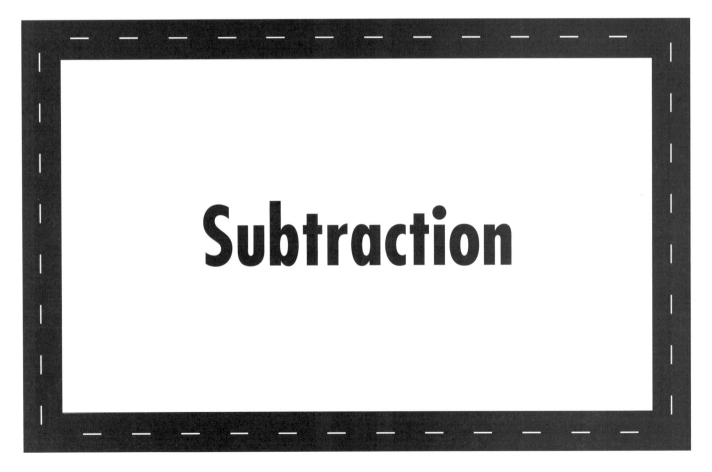

Operation Word Clues

Addition
In All
Together
Total
Sum
Perimeter

_____

_____

_____

Subtraction
Left   Difference
How many more   Begin with
How much more   Took Away
Fewer   Exceed

_____

_____

_____

Multiplication
Altogether
Product
How much
Times
Area

_____

_____

_____

_____

Division
Divided
How many for each
Quotient
Share
Distribute
Average

_____

_____

_____

Idea 12
Math Jeopardy

Playing Math Jeopardy will help students improve in their math skills, which will in turn help them improve in math problem solving. You can create your own Math Jeopardy game board by adding your own answers and associated question to the blank form provided, or you can use our example. We suggest that you use this game with small groups of students or, if using it with the whole class, split the class into groups of four to five students.

We have also provided a blank set of small cards for students to use for additional practice. Choose three to six categories in the skill areas you want to improve. Photocopy the cards so the answer is on one side and the question is on the back. The cards can be used for individual practice or used in groups.

Here's how to play.

❶ Photocopy your own Jeopardy statements and the answer key or the one provided onto an overhead transparency and project onto a whiteboard.

❷ Cover the answer key until a student answers a statement correctly, and then unveil the answer.

❸ Model by using the first one as an example. Read the statement and then ask your students to supply the question that goes with the answer.

❹ Choose students by calling names randomly, using noisemakers and calling on the first student to respond, or calling on students in order around the room.

❺ Each time a student or team of students provides the correct question in answer to a statement, put his or her name or the name of the group on the square.

❻ The student or group of students with the most points wins the game.

Math Jeopardy

What is the question?

OPERATIONS	METRIC SYSTEM	POLYGONS	NUMBER SYSTEMS
100 Points $+$	**100 Points** meter	**100 Points** square	**100 Points** $\{0, 1, 2, 3, \ldots\}$
200 Points quotient	**200 Points** base 10	**200 Points** 4-sided polygon	**200 Points** $<$
300 Points parentheses, exponents, multipliation and division, addition and subtraction	**300 Points** liter	**300 Points** equilateral	**300 Points** tenths
400 Points $a + b = b + a$	**400 Points** km	**400 Points** $a^2 + b^2 = c^2$	**400 Points** XCIV
500 Points $5!$	**500 Points** 1000 millimeters	**500 Points** tessellation	**500 Points** decimal number system

© 2006 by PRO-ED, Inc.

Idea 12

Math Jeopardy

Answer Key

OPERATIONS	METRIC SYSTEM	POLYGONS	NUMBER SYSTEMS
100 Points What is an addition sign?	**100 Points** What is main unit used to measure length in the metric system?	**100 Points** What is a rectangle with four congruent sides?	**100 Points** What are whole numbers?
200 Points What is the answer to a division problem?	**200 Points** What is the number base of the metric system?	**200 Points** What is a quadrilateral?	**200 Points** What is the symbol for less than?
300 Points What is the order of operations?	**300 Points** What is the main unit used to measure volume in the metric system?	**300 Points** What is a polygon called that has all congruent sides?	**300 Points** What place value is the digit to the right of a decimal point?
400 Points What is the commutative property of addition?	**400 Points** What is the abbreviation of kilometer?	**400 Points** What is the Pythagorean theorem?	**400 Points** What is 94?
500 Points What is factorial notation or $5 \times 4 \times 3 \times 2 \times 1$?	**500 Points** What is how many millimeters in a meter?	**500 Points** What is covering a plane with nonoverlapping polygons?	**500 Points** What is the number system that uses the digits 0 through 9, with 0 having the least value and 9 having the greatest value?

© 2006 by PRO-ED, Inc.

Idea 12

Math Jeopardy

What is the question?

100 Points	100 Points	100 Points	100 Points
200 Points	200 Points	200 Points	200 Points
300 Points	300 Points	300 Points	300 Points
400 Points	400 Points	400 Points	400 Points
500 Points	500 Points	500 Points	500 Points

Math Jeopardy

Answer Key

100 Points	100 Points	100 Points	100 Points
200 Points	200 Points	200 Points	200 Points
300 Points	300 Points	300 Points	300 Points
400 Points	400 Points	400 Points	400 Points
500 Points	500 Points	500 Points	500 Points

© 2006 by PRO-ED, Inc. Idea 12

Answer.	Answer.
Answer.	Answer.
Answer.	Answer.
Answer.	Answer.

Question?	Question?
Question?	Question?
Question?	Question?
Question?	Question?

© 2006 by PRO-ED, Inc. Idea 12

Idea 13
Highlighting

Completing math word problems involves a number of steps. Two important steps are identifying and using relevant information and identifying and ignoring irrelevant information. The process described in this idea is an easy-to-use technique that students can use when trying to complete math word problems. The strategy requires students to mark the material they are using; therefore, if the problem is in a textbook, make a copy of the page for the student to mark on.

Here's how Highlighting works.

❶ Tell students that you are going to show them a way to identify important information contained in the problems they read.

❷ Provide students with highlighters of the following colors: yellow, blue, pink, and purple.

❸ Give students a copy of the Highlighting Guide form, provided for this idea.

❹ Go through each step of the guide by modeling the steps.

❺ Complete as many examples as needed until students master the strategy and do not require your assistance.

Note: The color of the highlighters associated with the different features of the problem can be changed. Note that, when completing the second pass (highlighting the numbers that are needed), the color needs to change to a clearly different color.

Highlighting Guide

❶ Highlight **ALL NUMBERS** with the **YELLOW** highlighter.

❷ Highlight **ALL NUMBERS THAT YOU NEED** for the problem with the **BLUE** highlighter.

✻ *Use only the numbers that are now GREEN in color for solving the problem.*

❸ Highlight **ALL KEY WORDS** with the **PINK** highlighter.

❹ Highlight **THE KEY WORD** that will be used for labeling the answer with the **PURPLE** highlighter.

Highlighting Guide

❶ Highlight **ALL NUMBERS** with the **YELLOW** highlighter.

❷ Highlight **ALL NUMBERS THAT YOU NEED** for the problem with the **BLUE** highlighter.

✻ *Use only the numbers that are now GREEN in color for solving the problem.*

❸ Highlight **ALL KEY WORDS** with the **PINK** highlighter.

❹ Highlight **THE KEY WORD** that will be used for labeling the answer with the **PURPLE** highlighter.

Highlighting Guide

❶ Highlight **ALL NUMBERS** with the **YELLOW** highlighter.

❷ Highlight **ALL NUMBERS THAT YOU NEED** for the problem with the **BLUE** highlighter.

✻ *Use only the numbers that are now GREEN in color for solving the problem.*

❸ Highlight **ALL KEY WORDS** with the **PINK** highlighter.

❹ Highlight **THE KEY WORD** that will be used for labeling the answer with the **PURPLE** highlighter.

Highlighting Guide

❶ Highlight **ALL NUMBERS** with the **YELLOW** highlighter.

❷ Highlight **ALL NUMBERS THAT YOU NEED** for the problem with the **BLUE** highlighter.

✻ *Use only the numbers that are now GREEN in color for solving the problem.*

❸ Highlight **ALL KEY WORDS** with the **PINK** highlighter.

❹ Highlight **THE KEY WORD** that will be used for labeling the answer with the **PURPLE** highlighter.

© 2006 by PRO-ED, Inc.

Idea 13

Idea 14
What's Missing?

Students must be able to pick out the important information in math problems in order to solve them. This idea provides a twist that will help students understand what information is important in order to solve problems. An example and a blank What's Missing? form are provided.

Here's how it works.

❶ Develop several problems that have an important piece of information missing, or copy several problems from the text or curriculum guide, striking through one or two sentences that contain important information.

❷ Write two to four statements, one of which is the needed information. Vary the number and complexity of statements according to the level of the student you are teaching.

❸ Place students in small groups and have each group circle the letter of the statement that contains the information that is needed in order to solve the problem.

❹ Have each small group provide the missing information (answers will vary) and use their made up information to solve each problem.

❺ Have each group present one of the problems to the rest of the class. Make sure that they can explain how they got the answer.

Tip:

Provide to small groups of students several sets of four problems that have important information missing. Ask students to write three to four statements under each problem, with one statement containing the needed information. Photocopy each group's problems, and include them in a center for extra practice.

Note. This idea is based on *Mathematics for the Mildly Handicapped: A Guide to Curriculum and Instruction* (pp. 180–181), by J. F. Cawley, A. M. Fitzmaurice-Hayes, and R. A. Shaw, 1988, Boston: Allyn & Bacon. Copyright 1988 by Allyn & Bacon.

What's Missing?

Read each problem and decide what additional information you need to solve the problem. Below each problem are four statements. Circle the letter of the statement that describes the missing information. Next, provide the missing information (you will have to make it up), and use it to solve each problem. Choose one problem to present to the rest of the class.

1. A gardener is planting a garden with green peppers, broccoli, sweet potatoes, and tomatoes. She has 15 tomato plants, 6 broccoli plants, and 12 sweet potato plants. How many plants did she have altogether?

A The time of year the gardener is planting her garden

B The number of plants the gardener has already planted

C The size of the garden

D The number of green pepper plants the gardener has

2. A bowl of punch is made of pineapple, strawberry, and cranberry juices. The punch contains 1.5 quarts of fruit juice in all. If 0.62 quart of it is pineapple juice, how much cranberry juice is in the punch?

A The number of people who will drink the punch

B The amount of strawberry juice in the punch

C The number of cups in the punch

D The size of the punch bowl

3. Twenty-four students are to go on a field trip. How many cars are needed to take the children?

A The number of children that can go in each car

B The number of adults going on the field trip

C The distance to the field trip

D The day of the week the children are going on the field trip

4. David is saving to buy a new shirt. He has saved for 2 months. Last month he saved $25.00. How much did he save this month?

A The price of the shirt

B The color of the shirt

C The total amount saved

D The jobs David did to earn money

© 2006 by PRO-ED, Inc.

Idea 14

What's Missing?

Read each problem and decide what additional information you need to solve the problem. Below each problem are four statements. Circle the letter of the statement that describes the missing information. Next, provide the missing information (you will have to make it up), and use it to solve each problem. Choose one problem to present to the rest of the class.

❶ _____

A _____
B _____
C _____
D _____

❷ _____

A _____
B _____
C _____
D _____

❸ _____

A _____
B _____
C _____
D _____

❹ _____

A _____
B _____
C _____
D _____

Idea 15
What's the Question?

The strategy in this idea provides students with the opportunity to think about the problem-solving process and become teachers themselves as they construct the task. In this idea we teach students to select or write the best question for a problem. Once students are successful with selecting and writing good questions, you can use Idea 16, Can't Tell, to teach them to decide if they have enough information to answer the question asked.

Here are the steps to What's the Question?.

1. Present a sample problem (or use the samples provided) and three to four questions about it. Explain to the students that they will not have to answer the problem but rather they are to select (or write) the question the problem is asking. Ask students to select the question from the choices.

2. Use guided practice with additional word problems followed by three to four questions for practice.

3. After students are comfortable with the process, assign students to small groups and provide one card for each group member.

4. Have the students write a problem on the front of the card relating to the math unit currently being studied. On the back of the card, ask each student to write three to four questions about the problem.

5. Each group member will take turns presenting his or her problem to the group. The group will select one problem to present to the class.

Tip:
Use the completed cards in centers or as additional practice for small groups by having students check the box that depicts the best question.

Note. This idea is based on *Mathematics for the Mildly Handicapped: A Guide to Curriculum and Instruction* (p. 188), by J. F. Cawley, A. M. Fitzmaurice-Hayes, and R. A. Shaw, 1988, Boston: Allyn & Bacon. Copyright 1988 by Allyn & Bacon.

Examples

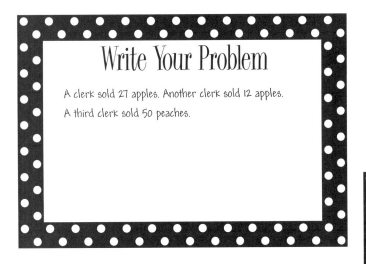

Write Your Problem

A clerk sold 27 apples. Another clerk sold 12 apples. A third clerk sold 50 peaches.

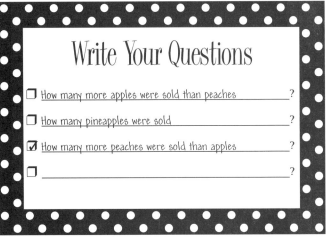

Write Your Questions

☐ How many more apples were sold than peaches ?
☐ How many pineapples were sold ?
☑ How many more peaches were sold than apples ?
☐ _____ ?

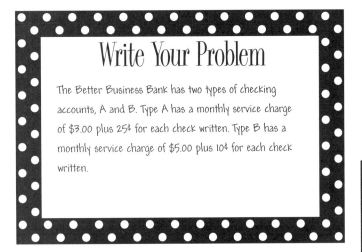

Write Your Problem

The Better Business Bank has two types of checking accounts, A and B. Type A has a monthly service charge of $3.00 plus 25¢ for each check written. Type B has a monthly service charge of $5.00 plus 10¢ for each check written.

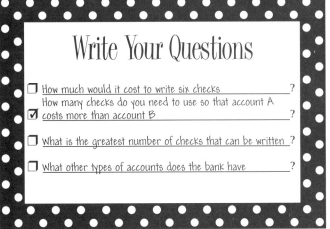

Write Your Questions

☐ How much would it cost to write six checks ?
☑ How many checks do you need to use so that account A costs more than account B ?
☐ What is the greatest number of checks that can be written ?
☐ What other types of accounts does the bank have ?

Idea 15

Idea 16
Can't Tell

Most problem-solving strategies emphasize finding the important information, selecting the operation, and determining the answer. Idea 15, What's the Question?, presented a method to teach students how to determine what question a math problem is asking. This idea provides a method to teach students to determine if they have enough information to answer the question.

Here are the steps to Can't Tell.

❶ Present a problem followed by three statements. One statement should be true, one false, and one should be unanswerable with the information provided in the problem. Tell students that if the problem contains insufficient information to determine if the answer is true or false, they should select the response Can't Tell. For example, our sample problem contains insufficient information to determine whether the statement "The truck was in front of the store" is true or false. Therefore, students would check Can't Tell.

❷ Provide students with additional word problems from the text or curriculum guide, followed by three statements for practice. Use the form provided.

❸ Have students perform the computation for those statements that are answered true or false, but not for statements marked as Can't Tell.

❹ Discuss how individuals must operate when using real-life math and how sometimes you must find the missing information before problems can be solved.

Judy delivered six coats from her truck to the store. Renee delivered three dresses from her truck to the store.

❶ Page __43__   Problem __6__

	TRUE	FALSE	CAN'T TELL
• The truck was in front of the store.	☐	☐	☑
• Together, they delivered nine pieces of clothing.	☑	☐	☐
• They delivered more dresses than coats.	☐	☑	☐

6 + 3 = 9
6 coats > 3 dresses

Note. **This idea is based on** *Mathematics for the Mildly Handicapped: A Guide to Curriculum and Instruction* (pp. 180–181), by J. F. Cawley, A. M. Fitzmaurice-Hayes, and R. A. Shaw, 1988, Boston: Allyn & Bacon. Copyright 1988 by Allyn & Bacon.

Can't Tell

Read the problem found on the page number provided. Read each statement and check the box that shows whether the statement is true, false, or if you can't tell. Prove the true and false statements in the space provided.

❶ Page _____ Problem _____

	TRUE	FALSE	CAN'T TELL
• _____	☐	☐	☐
• _____	☐	☐	☐
• _____	☐	☐	☐

❷ Page _____ Problem _____

	TRUE	FALSE	CAN'T TELL
• _____	☐	☐	☐
• _____	☐	☐	☐
• _____	☐	☐	☐

❸ Page _____ Problem _____

	TRUE	FALSE	CAN'T TELL
• _____	☐	☐	☐
• _____	☐	☐	☐
• _____	☐	☐	☐

❹ Page _____ Problem _____

	TRUE	FALSE	CAN'T TELL
• _____	☐	☐	☐
• _____	☐	☐	☐
• _____	☐	☐	☐

© 2006 by PRO-ED, Inc. Idea 16

Idea 17
Get Graphic

Mathematical information occurs in many forms. Important quantitative data are often presented in tables, figures, diagrams, and other graphically oriented formats. Students must be able to interpret relevant information from these formats. This idea assists students in pulling relevant information from any type of material that presents mathematical data in formats other than text. We have provided two forms: one is used when you want students to pull out the literal information, and the other is used when you want students to analyze the information.

Here's how it works.

❶ Direct students to examine the table, figure, diagram, or other graphic-oriented material.

❷ Teach students how to use Describe the Graphic by giving each student a copy of the form and following the directions below.

❸ After students understand how Describe the Graphic works, teach them how to use Interpret the Graphic by giving each student a copy of the form and following the directions below.

❹ Model how to complete the forms.
- Write the title and page number of the graphic on the first line.
- Circle the format of the graphic.

Describe the Graphic
- Ask students to identify the key labels used in the graphic and write them under the word labels.
- Record what each label represents next to it (e.g., name, probability).

Interpret the Graphic
- Write questions about the graph in the "Questions" column. The teacher can write these ahead of time or ask students to write their own.
- Have students answer each question in the "Answers" column.

Tip:
For homework, have students find a graph from a newspaper or news magazine, and complete the two forms.

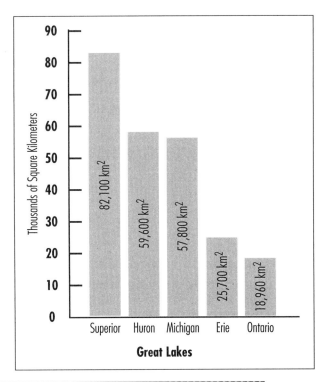

Describe the Graphic

Title of Graphic: Great Lakes   Page: 56

Format: table   graph   (diagram)   other

Label	What It Represents
• Great Lakes	• names of the five Great Lakes
• Square KM	• surface area in square km of each Great Lake
•	•

Write one fact about the graph.

Lake Superior is the largest.

Interpret the Graphic

Title of Graphic: Great Lakes   Page: 56

Format: table   graph   (diagram)   other

Questions	Answers
• Would it take longer to go around Lake Superior or Lake Michigan?	• Lake Superior
• Which lake is the smallest?	• Lake Ontario
• Which two lakes are about the same size?	• Lake Huron and Lake Michigan
•	•

Describe the Graphic

Title of Graphic: _____ **Page:** _____

Format: table   graph   diagram   other

Label	What It Represents
• _____	• _____
• _____	• _____
• _____	• _____

Write one fact about the graph.

Interpret the Graphic

Title of Graphic: _____ **Page:** _____

Format: table   graph   diagram   other

Questions	Answers
• _____	• _____
• _____	• _____
• _____	• _____
• _____	• _____

Idea 18
Use Manipulatives

Students learn best when teachers use a variety of instructional strategies, including hands-on strategies. Problem-solving tasks that use manipulatives provide many students with an entry into mathematics that they might not otherwise experience. When using manipulatives, it is important to follow some simple rules:

- Provide time at the beginning for free exploration. This is important so that students can satisfy their curiosity before tackling an assignment.
- Explain to students that there is a difference between playing with toys or games and using manipulatives.
- Set up a system for storing manipulatives and teach it to students.

Manipulatives can be purchased, homemade, or virtual. In this idea we illustrate how to use algebra tiles (along with a mnemonic for using them) and suggest several Web sites to learn more about using algebra tiles or using virtual manipulatives.

Here's how to use algebra tiles.

1. Purchase algebra tiles or use the patterns found at the end of this idea. If you use the patterns provided, color one side of the tiles to depict negative numbers.

2. Teach students the specific steps of STAR (see steps outlined in Idea 5, Mnemonics) if this approach will facilitate student understanding.

3. Teach students to use algebra tiles by following these steps (or follow steps outlined in your curriculum).
   a. Name the big square and the rectangle. For example, you can name them X^2 and X. The small squares represent integers and are countable.

Note. STAR is adapted from "Effects of a Problem-Solving Strategy on the Introductory Algebra Performance of Secondary Students with Learning Disabilities," by P. Maccini and C. A. Hughes, 2000, *Learning Disabilities Research & Practice, 15,* pp. 10–21. Copyright 2000 by the Division for Learning Disabilities. Adapted with permission.

b. Begin to work with the tiles by modeling how to name a collection of tiles. For example, the collection below would be named $2X^2 + X + 2$.

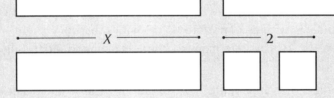

c. Once students are comfortable naming collections of tiles, you can begin to add and subtract collections and introduce the term *polynomials*.

d. When students have mastered addition and subtraction, begin to multiply and divide polynomials.

Additional Information on Algebra Tiles

Teaching with Algebra Tiles
http://plato.acadiau.ca/courses/educ/reid/Virtual-manipulatives/tiles/tiles.html

Algebra Tile Basics
www.kencole.org/algtilebasics.html

Purchase Algebra Tiles

Summit Learning Product Search
www.summitlearning.com/ent/search.cfm

Enasco Online Catalogs
www.enasco.com

Additional Information on Virtual Manipulatives

National Library of Virtual Manipulatives
http://nlvm.usu.edu/en/nav/index.html

Computing Technology for Math Excellence
www.ct4me.net/math_manipulatives.htm

Math: Virtual Manipulatives
www.emints.org/ethemes/resources/S00000592.shtml

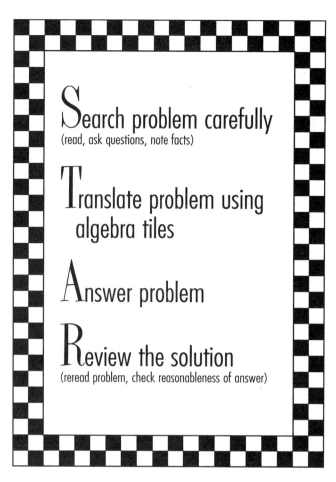
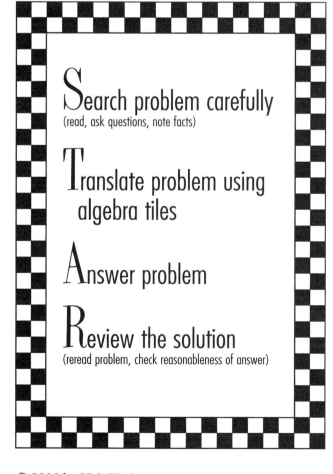

S T A R

To do the problem:

Search problem carefully ☐
(read, ask questions, note facts)

Translate problem using algebra tiles ☐

Answer problem ☐

Review the solution ☐
(reread problem, check reasonableness of answer)

Idea 19
Draw a Diagram

One basic problem-solving strategy that has helped students solve math problems for many years is drawing a diagram. The diagrams that students draw do not have to be perfect, as long as they are representative of the problem. Here are two examples and a form that students can use for the drawing, the number sentence, and the answer.

After you have taught students how to use Draw a Diagram, divide them into pairs and give each pair a new problem and one of the forms provided (i.e., use the form that matches the problem). Once they have solved it, allow each pair to demonstrate the process to the rest of the class, either working on the board or from a transparency.

> **Tip:**
> Ask students to make up additional questions that can be answered with the pictures.

Example Problem 1

Bobby has 4 cookies, his friend Mike has 1 cookie, and Hal has 5 cookies. What fraction of the total does each boy have?

① Project the problem. Show your students how to draw a diagram that represents the problem by drawing a row of circles representing the number of cookies for each boy. Bobby would have 4 circles, Mike 1 circle, and Hal 5 circles.

② Label the diagram by placing each boy's name under his row of cookies and the number of cookies under the name.

③ Write an equation or number sentence that reflects each part of the problem. Remind students that a fraction consists of two parts: the bottom part (denominator) is the total amount and the top part (numerator) is the part in which we are interested. It is best to begin by finding the total amount, which in this problem is the total number of cookies (4 + 1 + 5 = 10).

④ Finish the problem by finding the numerator of the fraction. For Bobby, the part we are interested in is 4 cookies, so place the 4 (part) over 10 (total). Do this for each person in the problem, reducing the fractions if possible.

THE PICTURE

○ ○ ○ ○
Bobby
4

○
Mike
1

○ ○ ○ ○ ○
Hal
5

THE PROBLEM

$4 + 1 + 5 = 10$

$\frac{4}{10} = \frac{2}{5}$

$\frac{1}{10}$

$\frac{5}{10} = \frac{1}{2}$

THE ANSWER

Bobby has $\frac{2}{5}$ of the cookies.

Mike has $\frac{1}{10}$ of the cookies.

Hal has $\frac{1}{2}$ of the cookies.

Example Problem 2

Four students were standing in line, waiting to go to lunch. Cindy was behind Beatriz. Jack was between Cindy and Beatriz. Cindy was in front of Omar. Who was in the back of the line?

① Draw a line to represent the lunch line. Write "front" and "back" to show the beginning and end of the line.

② Place two circles with the letters *C* and *B* in them on the line to represent the first two people, Cindy and Beatriz. Cindy should be behind Beatriz.

③ Tell the students you are going to add a third circle with the letter J in it to represent Jack. Ask the students where you should place it (between Cindy and Beatriz).

④ Tell students you are adding a fourth circle with an O in it to represent Omar. Ask students where you should place it (behind Cindy).

⑤ Ask students who was in the back of the line (Omar).

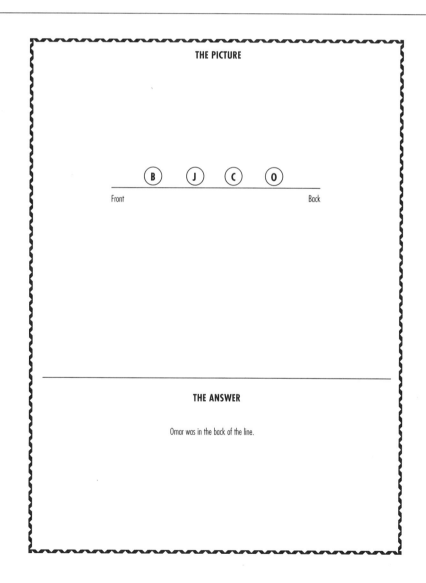

THE PICTURE

THE ANSWER

Omar was in the back of the line.

THE PICTURE

THE PROBLEM

THE ANSWER

THE PICTURE

THE ANSWER

Idea 20
Find a Pattern

Most of us look for patterns in what we read to help us make predictions about what will come next. In the same way, we can look for patterns in math problems to help us predict the answer. Some math problems require us to recognize a pattern, but other problems require us to extend a pattern to find a solution.

To teach students how to use the Find-a-Pattern strategy, model the strategy for them so that they know the steps. Next, use the strategy to solve several different problems, emphasizing the process. In this way, students can learn to recognize when to use the strategy. Finally, ask students to solve several problems, using the Find the Pattern steps.

Here are the steps to Find a Pattern.

❶ Read the problem and make a table to organize the data.

❷ Determine if the numbers are getting larger or smaller.

❸ Find the difference between successive pairs of numbers in the table.

❹ Use the difference to determine the pattern.

We have provided two examples to use when teaching students this strategy. The first example is for students in intermediate elementary grades, and the second is for students in prealgebra or algebra. A blank form is provided for each example so that you can make an overhead or use a document viewer to show the examples as you go through them.

Also provided are the Find-a-Pattern steps (a full page and small cards) that can be used as reminders of the steps and Congratulations cards that can be used as reinforcers. The Congratulations cards can be used with this idea and with Ideas 21 through 24. To use them, write in the student's name and the problem-solving strategy used.

Example Problem 1

Max's family bought a puppy. The veterinarian told Max that the puppy would gain about the same number of pounds each week for at least 6 weeks. The puppy weighed 18 lbs after the first week, 21 lbs after the second week, and 24 lbs after the third week. How much is the puppy likely to weigh after 6 weeks?

① *Read the problem and make a table to organize the data.*

Discuss what the data in the problem represent, in this case weeks and weight. Make a table with the weeks as column headers and the weight under the week it represents.

② *Determine if the numbers are getting larger or smaller.*

In this case the numbers are increasing, so we know the final answer must be larger than the beginning value.

③ *Find the difference between successive pairs of numbers in the table.*

Subtract each pair of numbers to find the difference of 3. The difference in this case represents the increase in weight.

④ *Use the difference to determine the pattern.*

Complete the table using the pattern.

Below is how the table should look.

Week 1	Week 2	Week 3	Week 4	Week 5	Week 6
18	21	24			
	−3	−3			
18	21	24	27	30	33

Example Problem 2

Leisha's family bought an ant farm with 1,500 ants. They were told that the farm could hold a total of 192,000 ants. On the second day the ant farm had 3,000 ants, and on the third day it had 6,000 ants. How many days will it take for the ant farm to be full?

① *Read the problem and make a table to organize the data.*

Discuss what the data in the problem represents, in this case, days and numbers of ants. Make a table with the days as column headers and the number of ants under the day it represents.

② *Determine if the numbers are getting larger or smaller.*

In this case the numbers are increasing, so we know the final answer must be larger than the beginning value.

③ *Find the difference between successive pairs of numbers in the table.*

Subtract each pair of numbers to find the difference of 1,500 and 3,000. The difference in this case doubles each day.

④ *Use the difference to determine the pattern.*

Complete the table using the pattern.

Below is how the table should look.

Day 1	Day 2	Day 3	Day 4	Day 5	Day 6	Day 7	Day 8
1,500	3,000	6,000					
	−1,500	−3,000					
1,500	3,000	6,000	12,000	24,000	48,000	96,000	192,000

Find a Pattern
Example Problem 1

Max's family bought a puppy. The veterinarian told Max that the puppy would gain about the same number of pounds each week for at least 6 weeks. The puppy weighed 18 lbs after the first week, 21 lbs after the second week, and 24 lbs after the third week. How much is the puppy likely to weigh after 6 weeks?

Week 1	Week 2	Week 3	Week 4	Week 5	Week 6

Find a Pattern
Example Problem 2

Leisha's family bought an ant farm with 1,500 ants. They were told that the farm could hold a total of 192,000 ants. On the second day the ant farm had 3,000 ants, and on the third day it had 6,000 ants. How many days will it take for the ant farm to be full?

Day 1	Day 2	Day 3	Day 4	Day 5	Day 6	Day 7	Day 8	Day 9	Day 10

Find a Pattern

Find a Pattern Steps

❶ Read the problem and make a table to organize the data in the problem.

❷ Determine if the numbers are getting larger or smaller.

❸ Find the difference between each pair of numbers in the table.

❹ Use the difference to determine the pattern.

Find a Pattern Steps

1. Read the problem and make a table to organize the data in the problem.
2. Determine if the numbers are getting larger or smaller.
3. Find the difference between each pair of numbers in the table.
4. Use the difference to determine the pattern.

Find a Pattern Steps

1. Read the problem and make a table to organize the data in the problem.
2. Determine if the numbers are getting larger or smaller.
3. Find the difference between each pair of numbers in the table.
4. Use the difference to determine the pattern.

Find a Pattern Steps

1. Read the problem and make a table to organize the data in the problem.
2. Determine if the numbers are getting larger or smaller.
3. Find the difference between each pair of numbers in the table.
4. Use the difference to determine the pattern.

Find a Pattern Steps

1. Read the problem and make a table to organize the data in the problem.
2. Determine if the numbers are getting larger or smaller.
3. Find the difference between each pair of numbers in the table.
4. Use the difference to determine the pattern.

© 2006 by PRO-ED, Inc.

Idea 20

Idea 21
Make a Table

Tables are powerful tools for organizing and understanding information found in math problems. This idea presents a technique to teach students how to use tables to make sense of the data contained in math problems.

To teach students how to use the Make a Table strategy, model the strategy for them so that they know the steps. Next, use the strategy to solve several different problems, emphasizing the process. In this way, students can learn to recognize when to use the strategy. Finally, ask your students to solve several problems using the Make a Table steps.

Here are the steps to Make a Table.

❶ Read the problem and write down the key words that refer to numerical information.

❷ Put each key word as the heading of a column.

❸ Organize the numerical information under each heading.

Tip:

Use the Congratulations cards found on page 117 as reinforcers. Write the student's name and Make a Table for the problem-solving strategy used.

Example Problem 1

You save $3 on Monday. Each day after that you save twice as much as you saved the day before. If this pattern continues, how much would you save in 5 days?

① *Read the problem, and write down the key words that refer to numerical information.*

days, money saved

② *Put each key word as the heading of a column.*

Discuss with the students how to begin with the known information, if possible. In this case, we know the number of days (i.e., 5) and use "Days" as the heading of Column 1. The heading of Column 2 should be "Amount Saved."

③ *Organize the numerical information under each heading.*

List days 1 through 5 under the "Days." We know we save $3 on Monday, or Day 1, and put it next to Day 1. Discuss with the students that the amount must be doubled each day. Do the calculations to complete the table.

Here is how the table should look.

Days	Amount Saved
1	$3
2	$6
3	$12
4	$24
5	$48

Answer: You would save $48 in 5 days.

Example Problem 2

Mr. Schultz owns a store and pays $4 for a box of 12 candy bars. He sells each candy bar for 50 cents. How much profit will he make if he sells 6 boxes of candy bars?

① *Read the problem, and write down the key words that refer to numerical information.*

boxes, profit

② *Put each key word as the heading of a column.*

Discuss with the students how to begin with the known information, if possible. In this case, we know the number of boxes (i.e., 6) and use "Boxes" as the heading of Column 1. The heading of Column 2 should be "Profit."

③ *Organize the numerical information under each heading.*

List boxes 1 through 6 under the heading in the first column. To answer this question, we first must figure out how much profit Mr. Schultz makes on the first day. If he sells the candy bars for 50 cents each, then he makes $6 on 12 candy bars. Because he buys the boxes for $4, he makes $2 profit. We will place $2 next to the number 1. Discuss with students that each box will earn the same amount of profit. Do the calculations to complete the table.

Here is how the table should look.

Boxes	Profit
1	$2
2	$4
3	$6
4	$8
5	$10
6	$12

Answer: Mr. Schultz would earn $12 profit on 6 boxes of candy bars.

Make a Table
Example Problem 1

You save $3 on Monday. Each day after that you save twice as much as you saved the day before. If this pattern continues, how much would you save in 5 days?

_____	_____

Make a Table
Example Problem 2

Mr. Schultz owns a store and pays $4 for a box of 12 candy bars. He sells each candy bar for 50 cents. How much profit will he make if he sells 6 boxes of candy bars?

_____	_____

Idea 21

Make a Table

_____	_____

Make a Table Steps

1 Read the problem, and write down the key words that refer to numerical information.

2 Put each key word as the heading of a column.

3 Organize the numerical information under each heading.

Make a Table Steps

1. Read the problem, and write down the key words that refer to numerical information.

2. Put each key word as the heading of a column.

3. Organize the numerical information under each heading.

Make a Table Steps

1. Read the problem, and write down the key words that refer to numerical information.

2. Put each key word as the heading of a column.

3. Organize the numerical information under each heading.

Make a Table Steps

1. Read the problem, and write down the key words that refer to numerical information.

2. Put each key word as the heading of a column.

3. Organize the numerical information under each heading.

Make a Table Steps

1. Read the problem, and write down the key words that refer to numerical information.

2. Put each key word as the heading of a column.

3. Organize the numerical information under each heading.

© 2006 by PRO-ED, Inc.

Idea 21

Idea 22
Make a List

Many of us make lists as a way to organize information. In the same way, lists can help us organize data in a math problem, which makes it easier to discover relationships and patterns in math problems.

To teach students how to use the Make a List strategy, model the strategy for them so that they know the steps. Next, use the strategy to solve several different problems, emphasizing the process. In this way, students can learn to recognize when to use the strategy. Finally, ask your students to solve several problems using the Make a List steps.

Here are the steps to Make a List.

❶ Read the problem, and write down the events that can occur.

❷ After each event, write the different ways it can occur.

❸ List one option from each event as your first combination.

❹ Continue until all possible combinations have been listed.

We have provided two examples to use when teaching students this strategy. The first example is for students in intermediate elementary grades, and the second example is for students in prealgebra or algebra. We suggest that you make an overhead or use a document viewer to show the examples as you go through them (see pages 130 and 131).

Tip:

Use the Congratulations cards found on page 117 as reinforcers. Write the student's name and Make a List for the problem-solving strategy used.

Example Problem 1

Customers at an ice-cream shop can choose from three flavors (chocolate, strawberry, vanilla) and two toppings (hot fudge, caramel) for their sundaes. How many one-flavor and one-topping sundaes are possible?

① *Read the problem, and write down the events that can occur.*
flavors, toppings

② *After each event, write the different ways it can occur.*
flavors—chocolate, strawberry, vanilla; toppings—hot fudge, caramel

③ *List one option from each event as your first combination.*
chocolate, hot fudge

④ *Continue until all possible combinations have been listed.*

Here is how the list should look.

chocolate	hot fudge
chocolate	carramel
strawberry	hot fudge
strawberry	caramel
vanilla	hot fudge
vanilla	caramel

Example Problem 1 has two events.

1. Three flavors
2. Two toppings

✶ Equation: $3 \times 2 = 6$

Tip:

Take these problems a step further by asking students to translate them into an equation using the sequential counting principle (i.e., event 1 × event 2 × event 3 . . .).

Example Problem 2

A spinner is divided into three equal parts, which are numbered 1, 2, and 3. A second spinner is divided into four equal parts, which are colored red, green, yellow, and blue. After the spinner comes to a stop, a pointer will be pointing at one of the numbers or colors. Ben and Shonda are playing a game in which they spin the first spinner, spin the second spinner, and flip a coin. How many possible outcomes of the first spinner, the second spinner, and the coin toss might occur?

① *Read the problem and write down the events that can occur.*
   spinner 1, spinner 2, coin toss

② *After each event, write the different ways it can occur.*
   spinner 1—1, 2, 3
   spinner 2—red, green, yellow, blue
   coin toss—heads, tails

③ *List one option from each event as your first combination.*
   1, red, heads

④ *Continue until all possible combinations have been listed.*

Here is how the list should look.

1, red, heads	2, red, heads	3, red, heads
1, green, heads	2, green, heads	3, green, heads
1, yellow, heads	2, yellow, heads	3, yellow, heads
1, blue, heads	2, blue, heads	3, blue, heads
1, red, tails	2, red, tails	3, red, tails
1, green, tails	2, green, tails	3, green, tails
1, yellow, tails	2, yellow, tails	3, yellow, tails
1, blue, tails	2, blue, tails	3, blue, tails

Example Problem 2 has three events.

1. Three outcomes: 1, 2, 3
2. Four outcomes: red, green, yellow, blue
3. Two outcomes: heads, tails

★ Equation: $3 \times 4 \times 2 = 24$

Make a List
Example Problem 1

Customers at an ice-cream shop can choose from three flavors and two toppings for their sundaes. How many one-flavor and one-topping sundaes are possible?

chocolate	hot fudge

Idea 22

Make a List
Example Problem 2

A spinner is divided into three equal parts, which are numbered 1, 2, and 3. A second spinner is divided into four equal parts, which are colored red, green, yellow, and blue. After the spinner comes to a stop, a pointer will be pointing at one of the numbers or colors. Ben and Shonda are playing a game in which they spin the first spinner, spin the second spinner, and flip a coin. How many possible outcomes of the first spinner, the second spinner, and the coin toss might occur?

1	red	heads

Idea 22

Make a List Steps

❶ Read the problem, and write down the events that can occur.

❷ After each event, write the different ways it can occur.

❸ List one option from each event as your first combination.

❹ Continue until all possible combinations have been listed.

Make a List Steps

1. Read the problem, and write down the events that can occur.
2. After each event, write the different ways it can occur.
3. List one option from each event as your first combination.
4. Continue until all possible combinations have been listed.

© 2006 by PRO-ED, Inc.

Idea 22

Idea 23
Work Backward

Sometimes it is helpful to start at the end of a problem and work back to the beginning. This strategy can be particularly helpful when you know the end goal and then work backward to determine how to meet the goal.

To teach students how to use the Work Backward strategy, model the strategy for them so that they know the steps. Next, use the strategy to solve several different problems, emphasizing the process so that students can learn to recognize when to use the strategy. Finally, ask your students to solve several problems using the Work Backward steps.

Here are the steps to Work Backward.

◆ Read the problem, and write down the end result.

◆ Work your way from the end of the problem to the beginning.

◆ At each step, do the necessary calculations.

We have provided two examples to use when teaching students this strategy. The first example is for students in intermediate elementary grades, and the second is for students in prealgebra or algebra. Make an overhead or use a document viewer to show the examples as you go through them (see pages 137 and 138).

Also provided are the Work Backward steps (a full page and small cards) that can be used as reminders of the steps and Congratulations cards that can be used as reinforcers.

Tip:

Use the Congratulations cards found on page 117 as reinforcers. Write the student's name and Work Backward for the problem-solving strategy used.

Example Problem 1

You went to an amusement park with a friend. You spent half of your money on admission to the park. You bought a souvenir with half the money that was left. Then you had $4 left, which you spent on lunch. How much money did you take to the amusement park?

① *Read the problem, and write down the end result.*
   The end result is that you have $0 left at the end of your day spent at the amusement park.

② *Work your way from the end of the problem to the beginning.*
   - At lunch time, the $4 left from the admission and purchase of the souvenir was spent.
   - Before lunch you spent half the money you had to purchase a souvenir. If you were left with $4 after purchasing the souvenir, and you used half to buy it, you must have had $8 before you bought the souvenir ($4 × 2 = $8).
   - The cost of admission was half of the amount that you brought for the day. After you spent the money on admission, you had $8 left. That means that admission was double this amount ($8 × 2 = $16). Therefore, you must have taken $16 to the amusement park.

③ *At each step, do the necessary calculations.*
   See step 2.

Example Problem 2

Ms. Holmes currently has 24 different kinds of cookies for sale in her bakery. A year after she went into business, she added 2 more kinds. Three years later, the number of different kinds of cookies tripled. How many different kinds of cookies did she sell in her bakery when she first opened?

① *Read the problem, and write down the end result.*
   The end result is that Ms. Holmes has 24 different kinds of cookies for sale.

② *Work your way from the end of the problem to the beginning.*
   - Three years after she opened up her business, Ms. Holmes tripled the number of different kinds of cookies she sold. Therefore, she had 8 different kinds of cookies before this (24 ÷ 3 = 8).
   - One year after Ms. Holmes opened her bakery, she added 2 more kinds of cookies. Therefore she began with 6 different kinds of cookies for sale in her bakery (8 − 2 = 6).

③ *At each step, do the necessary calculations.*
   See step 2.

Work Backward
Example Problem 1

You went to an amusement park with a friend. You spent half of your money on admission to the park. You bought a souvenir with half the money that was left. Then you had $4 left, which you spent on lunch. How much money did you take to the amusement park?

① Read the problem and write down the end result.

_____
_____

② Work your way from the end of the problem to the beginning.

- _____
_____
_____
_____

- _____
_____
_____
_____

- _____
_____
_____
_____

③ At each step, do the necessary calculations.

Work Backward
Example Problem 2

Ms. Holmes currently has 24 different kinds of cookies for sale in her bakery. A year after she went into business, she added 2 more kinds. Three years later, the number of different kinds of cookies tripled. How many different kinds of cookies did she sell in her bakery when she first opened?

① Read the problem and write down the end result.

② Work your way from the end of the problem to the beginning.

-
-
-

③ At each step, do the necessary calculations.

Work Backward

◆1 Read the problem, and write down the end result.

◆2 Work your way from the end of the problem to the beginning.

◆3 At each step, do the necessary calculations.

Work Backward

1. Read the problem, and write down the end result.
2. Work your way from the end of the problem to the beginning.
3. At each step, do the necessary calculations.

Work Backward

1. Read the problem, and write down the end result.
2. Work your way from the end of the problem to the beginning.
3. At each step, do the necessary calculations.

Work Backward

1. Read the problem, and write down the end result.
2. Work your way from the end of the problem to the beginning.
3. At each step, do the necessary calculations.

Work Backward

1. Read the problem, and write down the end result.
2. Work your way from the end of the problem to the beginning.
3. At each step, do the necessary calculations.

© 2006 by PRO-ED, Inc.   Idea 23

Idea 24
Guess and Test

Many of us make educated or "smart" guesses all the time. We base those guesses on our prior experience and knowledge. Although we may not always be conscious of the process we use to make smart guesses, our thinking probably follows some simple steps. Teaching your math students to make smart guesses is an excellent way to help them solve problems efficiently and correctly.

To teach students how to use the Guess and Test strategy, model the strategy for them so that they know the steps. Next, use the strategy to solve several different problems, emphasizing the process. In this way, students can learn to make smart guesses in a variety of situations. Finally, ask your students to talk themselves through a problem using the Guess and Test steps and see if it works for them.

Here are the steps to Guess and Test.

1. Read the problem, and make a smart guess based on the information.
2. Test your guess against the conditions of the problem.
3. Revise the guess.
4. Repeat until you get the correct answer.

We have provided two examples to use when teaching students this strategy. The first example is for students in intermediate elementary grades, and the second is for students in prealgebra or algebra. A blank form is provided for each example so that you can make an overhead or use a document viewer to show the examples as you go through them. Also provided are the Test and Guess steps (a full page and small cards) that can be used as reminders of the steps.

Tip:

Use the Congratulations cards found on page 117 as reinforcers. Write the student's name and Guess and Test for the problem-solving strategy used.

Example Problem 1

Shawna has 72¢. She has some pennies, nickels, and dimes. She has 10 coins in all. How many of each type of coin does she have?

① *Read the problem, and make a smart guess based on the information.*

Discuss why the best way to begin this problem is to figure out the number of pennies Shawna must have (i.e., 2 or 7). Use the blank table provided at the end of this idea, and complete it for the first guess the students make.

② *Test your guess against the conditions of the problem.*

Check the answer in the last column; if correct the problem is finished, if not proceed to step 3.

③ *Revise the guess.*

Make another guess; be sure students use the information from the previous to make a smart guess.

④ *Repeat until you get the correct answer.*

Continue the process, filling in the table until the students get the correct answer.

Below is one solution students may come up with.

Number of Pennies	Number of Nickels	Number of Dimes	Check	
7	1	2	7+5+20=32¢	(too small)
7	2	1	7+10+10=27¢	(too small)
2	4	4	2+20+40=62¢	(too small)
2	1	7	2+5+70=77¢	(too large)
2	2	6	2+10+60=72¢	(correct)

Example Problem 2

Three friends, Sam, Gerald, and Will, each have a different amount of 1-dollar bills. The total number of 1-dollar bills is 49. Sam has twice as many as Gerald. Will has twice as many as Sam. How many 1-dollar bills does each have?

① *Read the problem, and make a smart guess based on the information.*

Discuss why the best way to begin this problem is to figure out who has the fewest dollar bills. Use the blank table provided at the end of this idea, and complete it for the first guess the students make.

② *Test your guess against the conditions of the problem.*

Check the answer in the last column; if correct the problem is finished, if not proceed to step 3.

③ *Revise the guess.*

Make another guess; be sure students use the information from the previous guess to make a revised smart guess.

④ *Repeat until you get the correct answer.*

Continue the process, completing the table until the students get the correct answer.

Below is one solution students may come up with.

Sam	Gerald	Will	Check	
8	4	16	8+4+16=28	(too small)
20	10	40	20+10+40=70	(too large)
12	6	24	12+6+24=42	(too small)
14	7	28	14+7+28=49	(correct)

Tip:

To take this one problem one step further, ask students to translate it into an algebraic expression. Let Sam = s, Gerald = g, Will = w.

$s + g + w = 49$

$s = 2g$

$w = 2s$

$w = 2 \cdot 2g$

$w = 4g$

Substitute: $g + 2g + 4g = 49$

Guess and Test
Example Problem 1

Shawna has 72¢. She has some pennies, nickels, and dimes. She has 10 coins in all. How many of each type of coin does she have?

Number of Pennies	Number of Nickels	Number of Dimes	Check

Guess and Test
Example Problem 2

Three friends, Sam, Gerald, and Will, each have a different amount of 1-dollar bills. The total number of 1-dollar bills is 49. Sam has twice as many as Gerald. Will has twice as many as Sam. How many 1-dollar bills does each have?

Sam	Gerald	Will	Check

Idea 24

Guess and Test

_____	_____	_____	Check

Idea 24

Guess and Test Steps

❶ Read the problem, and make a smart guess based on the information.

❷ Test your guess against the conditions of the problem.

❸ Revise the guess.

❹ Repeat until you get the correct answer.

Idea 25
Estimation Games

There are many specific skills that students need to be effective problem solvers. Games are a fun way to reinforce these skills. This idea presents two games that can be used to improve students' skills and make them better problem solvers.

Estimate It

This game is a fun way for groups of two to four students to practice estimating by rounding. All you need is the set of cards numbered 0 through 9 for each player (provided at the end of this idea) and paper and pencils.

❶ One player shuffles the cards and deals four to each player. Leaving the cards dealt face down, each student arranges the cards in two columns (addition format) with two cards above and two cards below.

❷ Students turn over the two cards that are in the tens place, and keep the units cards face down.

❸ Each student estimates the sum of his or her cards and records it on a piece of paper.

❹ The cards in the units place are turned over, and the actual sum is computed. The difference between the estimate and the actual sum is found. This amount is recorded.

❺ The game ends when one player reaches 20 points. The player with the lowest score is the winner.

> **Tip:**
> Vary this game by dealing 6 or 8 cards to estimate to the nearest 100 or 1,000. Increase the number of points needed to end the game accordingly.

In Your Head

In Your Head is a great way for students to practice calculating in their heads, pick out the important information in a problem, and reinforce math vocabulary. To play, you need to make enough copies of the number cards (1 through 30) provided at the end of this idea so that each student has five cards and one copy of the question cards for the leader (a teacher or a student). You can use the blank question cards to create your own questions to focus on the operations that are at the level of your students.

1. Shuffle the cards and give each student 5 of the number cards.

2. Stack the question cards face down, and project them one at a time. Leave each card projected for 1 minute. Do not allow students to use paper and pencil.

3. Students who have the number card that answers the question turns it over.

4. The first student to turn over all his or her number cards is the winner. The leader checks the answers to make sure the student is correct.

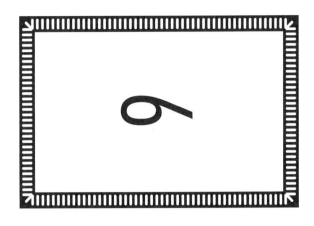

© 2006 by PRO-ED, Inc.

Idea 25

© 2006 by PRO-ED, Inc.

Idea 25

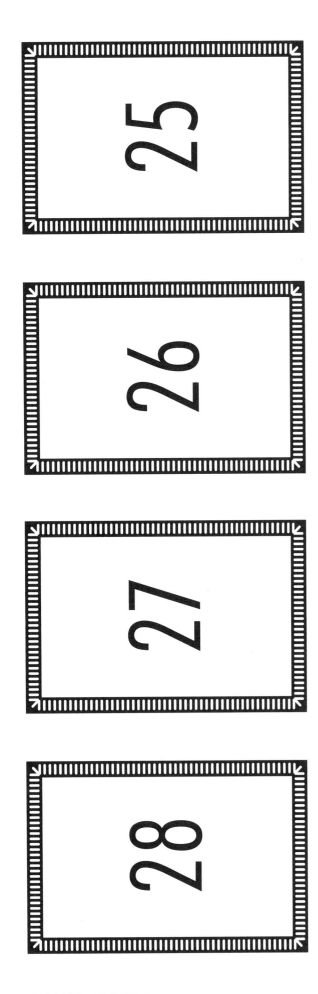
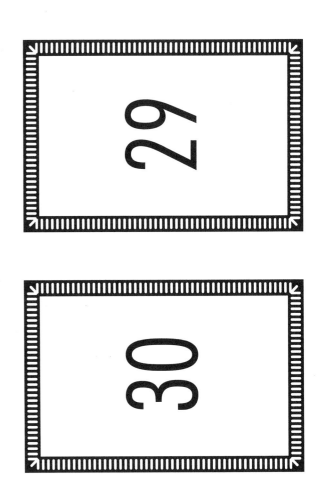

Idea 25

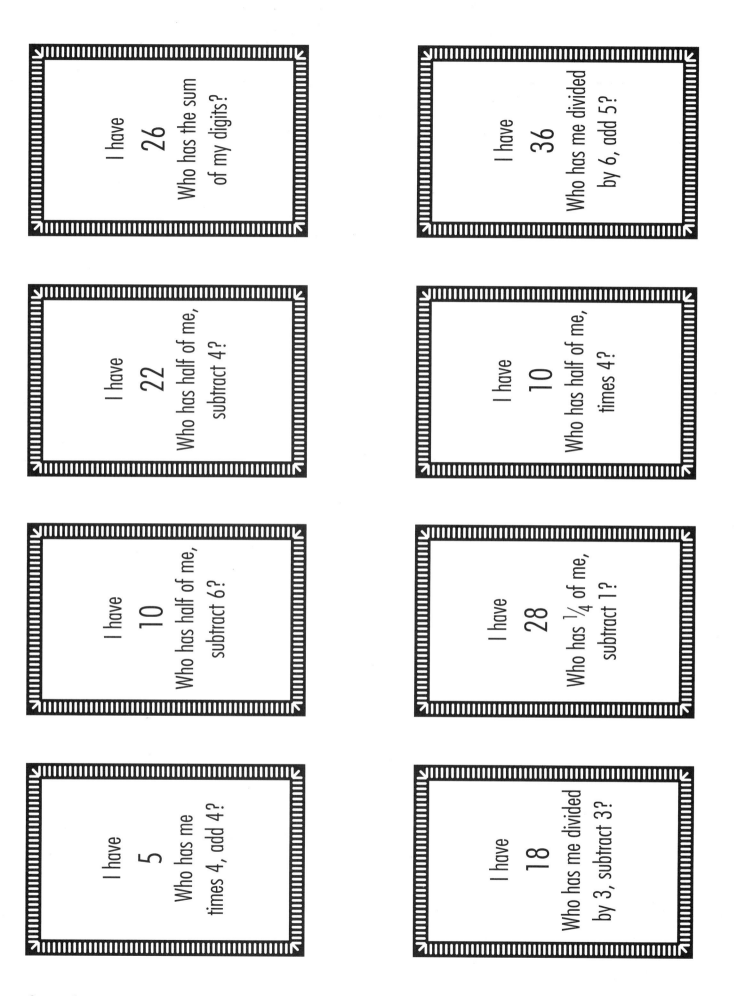

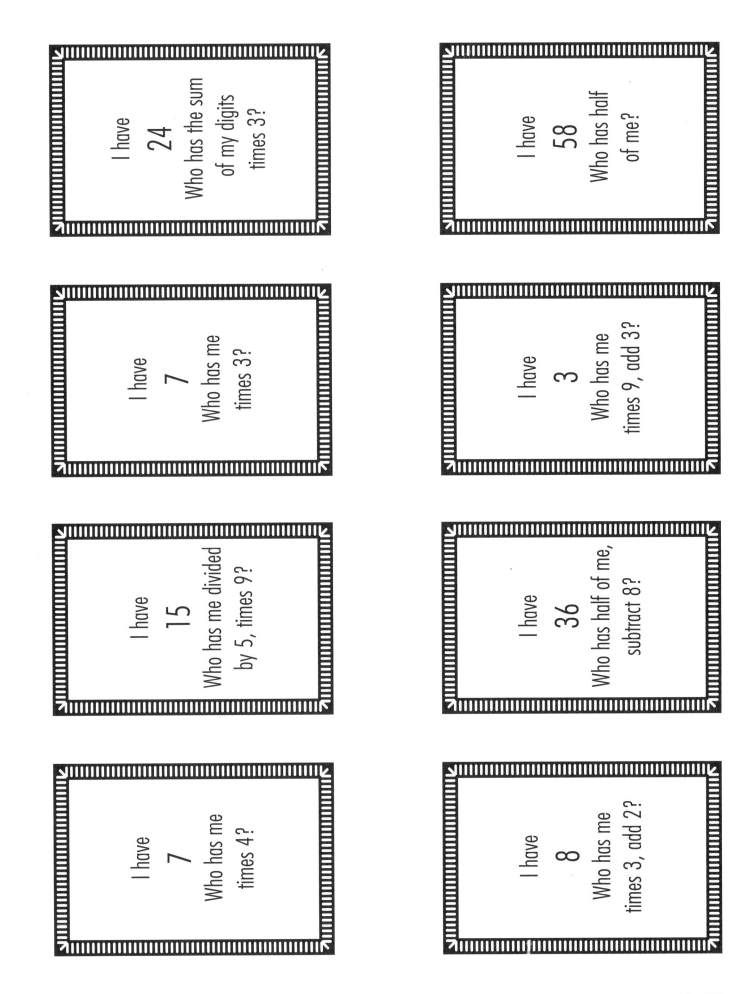

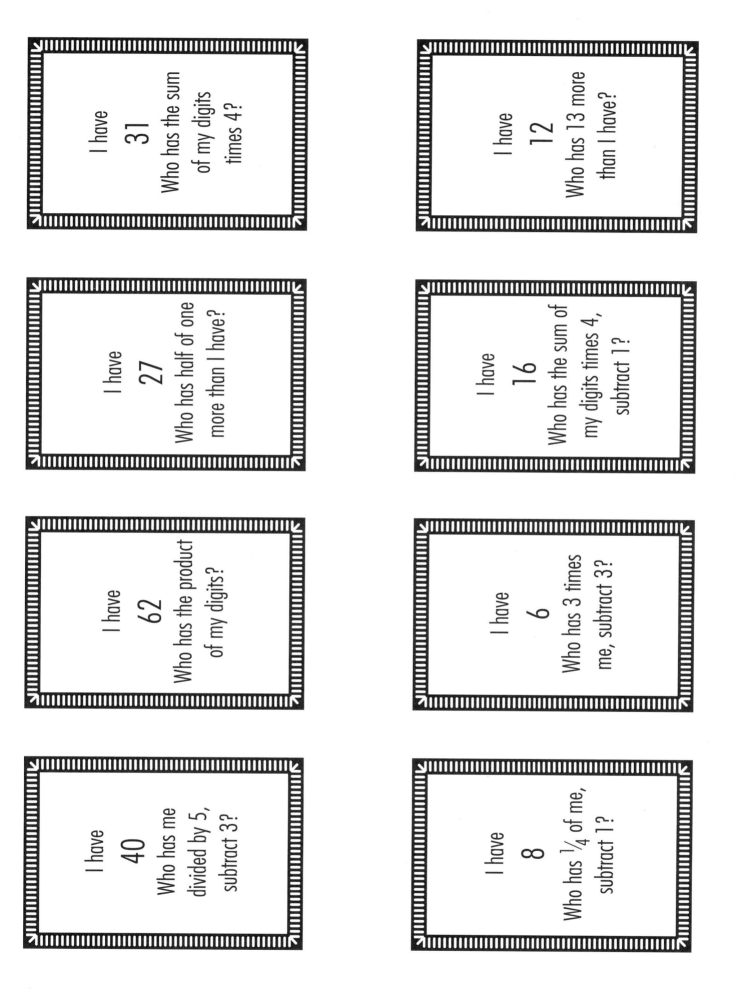

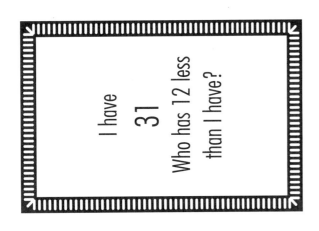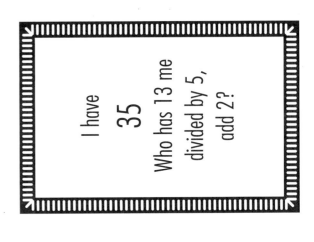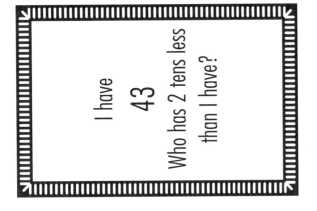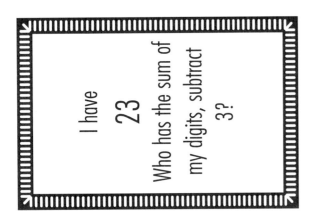

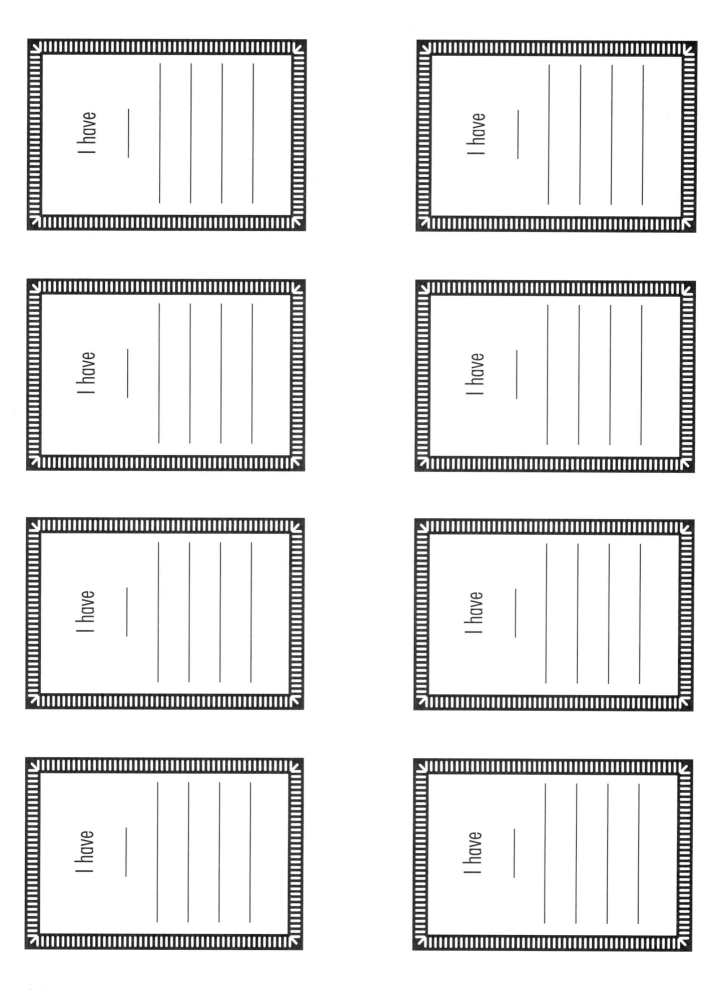

Idea 26
The Price Is Right

Estimation is a critical skill for math problem solving, and The Price Is Right is a fun way to reinforce this skill. Before playing the game, make sure that you have taught students the skill of estimation and, in particular, how to use rounding when estimating.

To play the Price Is Right, take these steps.

❶ Present several items that cost varying amounts (take digital photos or cut pictures from magazines).

❷ Tell students that they have a set amount of money. This amount will vary depending on the age and level of students you teach.

❸ Ask the students to pick out three items they would like to purchase, and list them on the blanks on the form provided.

❹ Set a timer for a predetermined amount of time, and tell students to decide if they have enough to buy their choices by estimation. Do not allow them to use paper and pencil.

❺ When the timer goes off, ask students to check the *yes* box if they have enough money to buy what they have chosen and the *no* box if they do not have enough money.

❻ Have the students check themselves by adding the prices of the three items together.

Variations

⊙ Choose three or more items for the entire class, and ask them to estimate the total cost in whole dollars. Use as a timed or untimed activity.

⊙ Divide the class into pairs. Have the first person in the pair choose three items and estimate the cost. The second person checks the answer. Switch roles and repeat.

⊙ Divide the class into teams of six or more. Give the first member on each team a list of products to estimate to the nearest whole dollar. The team member who provides the correct estimate first receives a point for his or her team. When all team members have had an opportunity to estimate, the team with the most points wins.

The Price Is Right

Name: _____

List your three product choices.

1._____
2._____
3._____

Do you have enough money? ❑ Yes ❑ No

Check yourself by adding the prices of your items in the space below.

Were you correct? ❑ Yes ❑ No

The Price Is Right

Name: _____

List your three product choices.

1._____
2._____
3._____

Do you have enough money? ❑ Yes ❑ No

Check yourself by adding the prices of your items in the space below.

Were you correct? ❑ Yes ❑ No

The Price Is Right

Name: _____

List your three product choices.

1._____
2._____
3._____

Do you have enough money? ❑ Yes ❑ No

Check yourself by adding the prices of your items in the space below.

Were you correct? ❑ Yes ❑ No

The Price Is Right

Name: _____

List your three product choices.

1._____
2._____
3._____

Do you have enough money? ❑ Yes ❑ No

Check yourself by adding the prices of your items in the space below.

Were you correct? ❑ Yes ❑ No

Idea 27
Beginning Problem Solving

Word problems are typically introduced in math textbooks at the end of the first grade and continue to be found in textbooks with increasing levels of difficulty and sophistication throughout the elementary grades. Initially, these problems involve the operations of addition and subtraction and eventually involve the use of multiplication and division.

The suggestions associated with this idea focus only on word problems that involve the operations of addition and subtraction. Idea 28, Which Operation?, provides a more advanced strategy for dealing not only with word problems that involve addition and subtraction but also with multiplication and division. Here, word problems involving addition and subtraction are categorized into the following three problem types:

- **Time-sequence problems** involve a sequence of actions or events in which something is added or taken away from an original situation and typically include verbs such as *gets, gives, finds, loses, buys, sells, eats,* and *drinks*.

- **Comparison problems** involve a comparison between two people or items in terms of some dimension, such as bigger/smaller, longer/shorter, older/younger, heavier/lighter, and so forth.

- **Classification problems** involve the breakdown of some type of material (i.e., a class) such as cars into more defined subcategories (e.g., red cars vs. blue cars).

> **Tip:**
> Each of the following strategies must be taught using explicit instruction that includes many examples. You are encouraged to see Stein et al. (2006) for detailed explanations on how to teach these strategies.

Note. This idea is based on instructional strategies from *Designing Effective Mathematics Instruction: A Direct Instruction Approach* (4th ed., pp. 207–212), by M. Stein, D. Kinder, J. Silbert, and D. W. Carnine, 2006, Upper Saddle River, NJ: Prentice Hall. Copyright 2006 by Prentice Hall.

Time-Sequence Problems

Time-sequence problems involve a series of events that change the quantity of the beginning event. The wording of the problem can be confusing for students, resulting in difficulty choosing the correct operation—addition or subtraction. Two example time-sequence problems, along with completed forms, are provided; the first is straightforward, and the second is linguistically difficult.

Example 1: Jackie has 5 colored pencils. Her mom gives her 2 more. How many pencils does she have now?

Example 2: Chris likes pets. Today he has 10 different pets. Yesterday he got 2 more. How many pets did he have last week?

Here's how to teach this strategy.

❶ Give each student a copy of the Time Sequence Form, found at the end of this idea. Teach students the vocabulary found at the top. Specifically,

   a. *Start with:* number that is at the beginning of the action or sequence of events

   b. *In:* any number that is added to the problem

   c. *Out:* any number that is removed from the problem

   d. *End with:* number that is at the end of the action or sequence of events

❷ Project a time-sequence word problem (we suggest that you begin with example 1, found on p. 167) and complete the top section with the students.

❸ Transfer the numbers from the top section to the appropriate row in the middle section by finding the row that has the two corresponding bulleted boxes. Each row in the middle section has either a + or − in front of it. This indicates what operation to use.

❹ Transfer the numbers to the bottom section, circle the operation that was indicated by the middle section, and compute the answer. Make sure to include a label. For addition, the numbers can be placed in either blank. For subtraction, teach the students to always place the larger number in the first blank.

❺ Do several problems with the students, teaching them how to use the form. Once students understand how to use it, assign several time-sequence problems to complete using the form.

Tip: The forms can be copied onto card stock or other sturdy paper and laminated so that the student can reuse the form and paper can be conserved.

Example Time-Sequence Problem 1

Jackie has 5 colored pencils. Her mom gives her 2 more. How many pencils does she now have?

Explanation

① In the top section, place the 5 in the "Start with" box because 5 is the number of pencils Jackie had in the beginning. The 2 is placed in the "In" box because Jackie is given 2 more pencils.

② Transfer the numbers from the top section to the first row in the middle section, because it contains the two corresponding bulleted boxes. The operation indicated is addition.

③ Transfer the numbers again to the bottom section, circle the addition sign, compute the answer, and fill in the label "pencils."

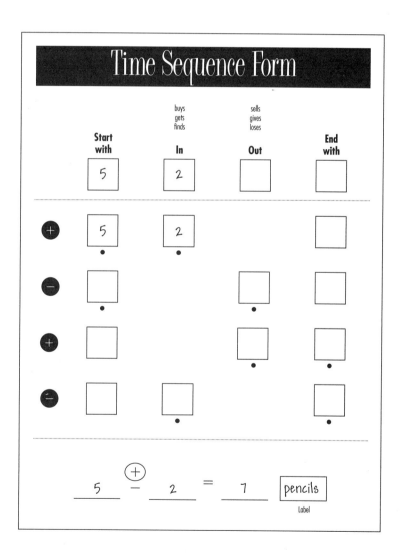

Example Time-Sequence Problem 2

Chris likes pets. Today he has 10 different pets. Yesterday, he got 2 more. How many pets did he have last week?

Explanation

① In the top section, place the 10 in the "End with" box because 10 is the number of pets Chris ends up with. The 2 is placed in the "Out" box because Chris didn't have these pets yesterday.

② Transfer the numbers from the top section to the last row in the middle section, because it contains the two corresponding bulleted boxes. The operation indicated is subtraction.

③ Transfer the numbers again to the bottom section, circle the subtraction sign, compute the answer, and fill in the label "pets." Make sure the students place the larger number, 10, in the first blank.

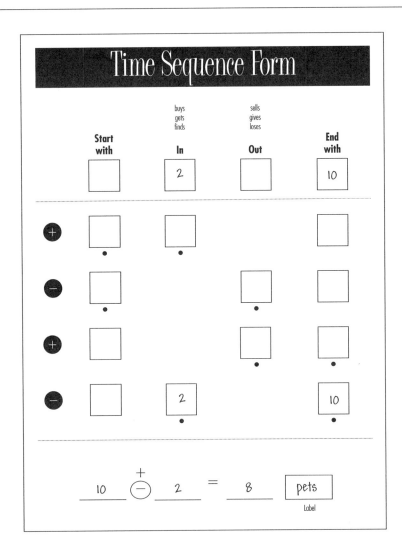

Comparison Problems

Comparison problems are exactly what the name indicates—these problems involve some type of comparison between two items or people. The characteristic feature of these problems will be the use of certain adjectives (e.g., bigger/smaller, longer/shorter, older/younger, heavier/lighter) that involve quantity. The following two examples illustrate comparison problems from relatively straightforward to linguistically difficult.

Example 1: Tom is 18 years old. Doug is 3 years older than Tom. How old is Doug?

Example 2: Glenn ran for 30 minutes. Glenn ran for 11 minutes longer than Dave did. How long did Dave run?

Many comparison problems are straightforward because the question simply asks the student to find the difference between two quantities. (e.g., Sylvia has two picture frames. The first one is 24 inches wide. The other frame is 15 inches wide. How much wider is the first picture frame?) This form is not for these types of problems.

Here's how to teach this strategy.

❶ Give each student a copy of the Comparison Form, found at the end of this idea.

❷ Project a comparison word problem (we suggest that you begin with Example 1, found on p. 170).

❸ In the middle box that is slightly elevated the student should enter the largest number given. A label (e.g., person name or item name) should be written on the line below the box.

❹ Ask the students to read the sentence that explains the comparison being made and to indicate the comparison quantity (number).

❺ Read the question to determine what we are being asked to provide. The comparison quantity will be written in either box A or B, depending on the following:
  - If the final result implies a greater quantity (e.g., larger, longer, older, etc.), the number should be inserted in box B.
  - If the final result implies a quantity of lesser quantity (e.g., smaller, shorter, younger, etc.), the number should be inserted in box A

❻ Identify which operation to use by observing the operation sign that is written above the box.

❼ Transfer the numbers to the bottom section, circling the operation indicated and compute the answer. Make sure to include a label.

❽ Do several problems with the students, teaching them how to use the form. Once students understand how to use it, assign several comparison problems to complete using the form.

Example Comparison Problem 1

Tom is 18 years old. Doug is 3 years older than Tom. How old is Doug?

Explanation

① The largest number given is 18, so place 18 in the middle box.

② The second sentence tells us that Doug is older than Tom. The comparison number is 3. Because Doug is older, the 3 is placed in box B, indicating addition.

③ Transfer the numbers to the bottom section, circle the addition sign, compute the answer, and fill in the label, "years old."

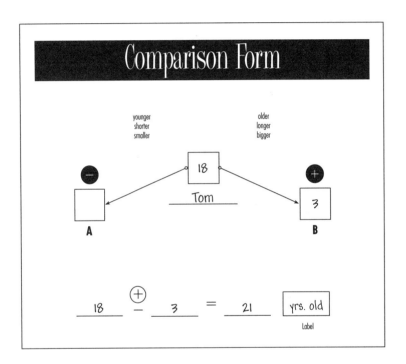

Example Comparison Problem 2

Glenn ran for 30 minutes. Glenn ran for 11 minutes longer than Dave did. How long did Dave run?

Explanation

① The largest number given is 30, so place 30 in the middle box.

② The second sentence tells us that Glenn ran longer than Dave. The comparison number is 11. The question asks how long Dave ran, and because Dave ran shorter (remember that Glenn ran longer), the 11 is placed in box A, indicating subtraction.

③ Transfer the numbers to the bottom section, circle the addition sign, compute the answer, and fill in the label, minutes.

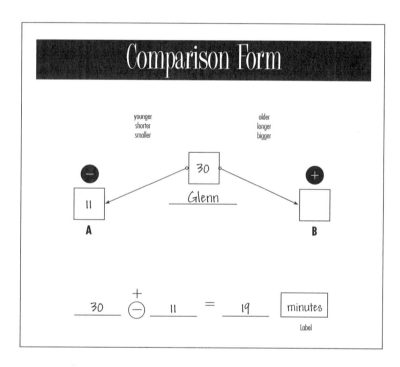

Classification Problems

Classification problems involve situations where classes (e.g., animals) and subclasses (e.g., dogs, cats) are presented. Students must be able to determine what information indicates classes or the total and what information indicates subclasses or types. Based on whether one has the total or not will determine which operation to use—addition or subtraction. The following two examples illustrate classification problems.

Example 1: Lyndsay had 8 pets at home: 5 were guinea pigs, the rest were rabbits. How many rabbits did she have?

Example 2: Steve played sports that used rackets. He has 6 tennis rackets and 5 racquetball rackets. How many rackets does he have in all?

Here's how to teach this strategy.

❶ Give each student a copy of the Classification Form, found at the end of this idea.

❷ Project a classification word problem (we suggest that you begin with example 1, found on p. 173).

❸ Teach students that the class or the total is placed in box C, and the subclasses or types are placed in boxes A and B. Read the first problem together and ask students what is the class or the total. Write this word on the line under box C. Ask students what are the subclasses and write these words on the lines above boxes A and B.

❹ Fill in the boxes according to the information provided in the problem. Based on which boxes are filled in, the student should do the following:

　a. If box C and either box A or B is filled in, then the operation that is needed is subtraction.

　b. If both boxes A or B are filled in and box C is empty, then the operation that is needed is addition.

❺ Transfer the numbers to the bottom section, circling the operation indicated, and compute the answer. Make sure to include a label. For addition, the numbers can be placed in either blank. For subtraction, teach the students to always place the larger number in the first blank.

❻ Do several problems with the students, teaching them how to use the form. Once students understand how to use it, assign several classification problems to complete using the form.

Example Classification Problem 1

Lyndsay had 8 pets at home, 5 were guinea pigs. The rest were rabbits. How many rabbits did she have?

Explanation

① The class or total in this problem is 8 pets, so 8 is placed in box C, and *pets* is written on the line underneath the box.

② Guinea pigs and rabbits are subclasses or types of pets, so write the names on the lines above boxes A and B. The number of guinea pigs is give, so place 5 in the box under *guinea pigs*. Because box A and box C are filled in, we subtract.

③ Transfer the numbers to the bottom section, circle the subtraction sign, compute the answer, and fill in the label, "rabbits."

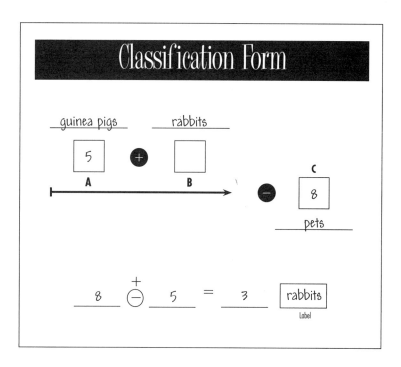

Example Classification Problem 2

Steve played sports that used rackets. He has 6 tennis rackets and 5 racquetball rackets. How many rackets does he have in all?

Explanation

① The class or total in this problem is rackets, so rackets is written on the line underneath box C.

② Tennis and racquetball are subclasses or types of rackets, so write the names on the lines above boxes A and B. The number of tennis rackets and racquetball rackets are given, so place the corresponding numbers in the boxes under each heading. Because box A and box B are filled in, we add.

③ Transfer the numbers to the bottom section, circle the addition sign, compute the answer, and fill in the label, "rackets."

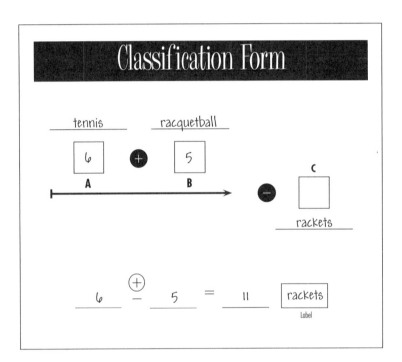

Time Sequence Form

	buys gets finds	sells gives loses	
Start with	**In**	**Out**	**End with**

$$\underline{\qquad} \begin{matrix}+\\-\end{matrix} \underline{\qquad} = \underline{\qquad} \boxed{}$$
Label

© 2006 by PRO-ED, Inc.

Idea 27

Comparison Form

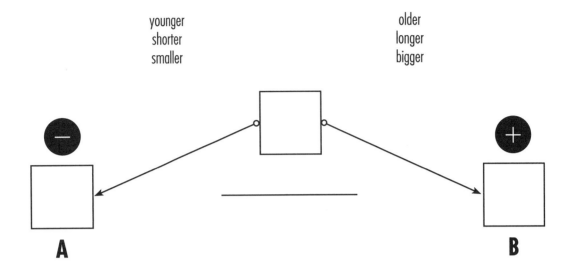

Classification Form

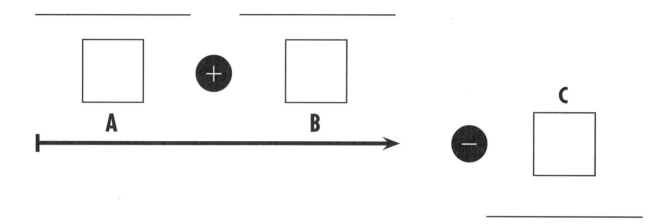

_____ +
_____ − = _____ [   ]
                    Label

Idea 27

Idea 28
Which Operation?

A key challenge for students solving math word problems is determining the correct operation to use in a given problem. This challenge has been addressed by using questions that use words (e.g., How many did the boys have *all together*?) that provide clues about which operation to use (in this example, addition). Idea 11, Word Clues, teaches students to look for clue words. However, problems in the real world frequently do not come with built-in cuing systems and, consequently, students need other methods for determining which operation to use.

Here's a method that helps students select the correct operation.

❶ Project the large card onto a whiteboard or other surface. Teach students the two key rules that they must follow in order to determine the correct operation.

Rule 1: Do you have the large number, or do you need to get the large number? If you have the large number already, subtract or divide. If you need to get the large number, add or multiply.

Rule 2: Are the two groups the same size or different sizes? If you have groups that are the same size, divide or multiply. If the groups are of different sizes, add or subtract.

❷ Present the example problems provided, or create your own to illustrate how the rules work.

❸ Once the students are comfortable with the process, give them the cue cards to keep in their math folders. Provide several problems to promote mastery and generalization.

Tip:
Assign students a set of problems that they do not have to solve. Instead, ask them to write the operation they should use next to the problem number.

Note. This graphic representation of math problem solving is classroom derived and is consistent with graphic representation strategies for solving word problems from "Teaching Middle School Students with Learning Disabilities To Solve Multistep Word Problems Using a Schema-Based Approach," by A. K. Jitendra, K. Hoff, and M. Beck, 1999, *Remedial and Special Education, 20*(1), pp. 50–64. Copyright 1999 by PRO-ED, Inc., and "Teaching Students Math Problem Solving Through Graphic Representations," by A. K. Jitendra, 2002, *Teaching Exceptional Children, 34*(4), pp. 34–38. Copyright 2002 by Council for Exceptional Children.

Which Operation?

	Have the Larger Number	Need the Larger Number
Same Size or Equal Groups	÷	×
Different Size or Unequal Groups	−	+

Which Operation?

❶ Miguel had 82 cents. He spent 65 cents on a soda. How much does he have?

	Have the Larger Number	Need the Larger Number
Same Size or Equal Groups	÷	×
Different Size or Unequal Groups	−	+

Explanation

Rule 1: We have the larger number, because 82 cents is the total. Miguel is going to spend some of the total, and the amount we are trying to find is going to be smaller than the total. Rule 1 tells us we must subtract or divide.

Rule 2: We have different amounts of money, because 82 and 65 are different sizes. Rule 2 tells us we must subtract.

❷ Sally read 12 textbook pages one day and 32 pages the next day. How many pages did she read?

	Have the Larger Number	Need the Larger Number
Same Size or Equal Groups	÷	×
Different Size or Unequal Groups	−	+

Explanation

Rule 1: We need the larger number because we have two different amounts of pages read on 2 days, and we want to know the total number of pages read. Rule 1 tells us we must add or multiply.

Rule 2: We have different numbers of pages, because 12 and 32 are different sizes. Rule 2 tells us we must add.

❸ Kimi had 4 boxes of cookies. There are 5 cookies in each box. How many cookies does she have?

	Have the Larger Number	Need the Larger Number
Same Size or Equal Groups	÷	×
Different Size or Unequal Groups	−	+

Explanation

Rule 1: We need the larger number, because we have 4 boxes of cookies and want to know the total number in all 4 boxes. Rule 1 tells us we must add or multiply.

Rule 2: We have the same number of cookies in each box, so we have equal groups of 5 cookies each. Rule 2 tells us we must multiply.

© 2006 by PRO-ED, Inc.

Idea 28

Which Operation?

❹ The boy scouts were going to go caving. Each of the scouts had flashlights and had to carry extra batteries. The scout leader gave them 40 batteries and wanted each scout to carry an equal amount. How many batteries does each boy scout have to carry?

	Have the Larger Number	Need the Larger Number
Same Size or Equal Groups	÷	×
Different Size or Unequal Groups	−	+

Explanation

Rule 1: We have the larger number, in this case it is 40 batteries. These need to be distributed to each scout. Rule 1 tells us that we must subtract or divide.

Rule 2: We have one group of the same size, 5 scouts. Rule 2 tells us we must divide.

Idea 29
Formulas

Formulas are powerful mathematical tools. Many common forms of word problems that use formulas occur with such frequency that it is important to teach the formula to students. This will ensure that they can successfully attack these problems. This idea provides a method for teaching three common formula problems.

Here's how it works.

1. Provide examples of problems for each type of formula, and highlight the kinds of questions that are commonly asked.

2. Introduce the formula by giving each student the card or sheet found at the end of this idea to place in his or her notebooks.

3. Model and demonstrate how to use the formula with several sample problems.

4. Use guided practice and subsequently independent practice to ensure that students understand and can apply the strategy.

 **Tip:**
Allow students to use the cards when completing tests.

Probability

$$\text{probability of an event} = \frac{\text{number of favorable outcomes}}{\text{number of possible outcomes}}$$

A common task in problem solving requires determining the probability of an event. Probability is an important tool in the solution of many problems in everyday life (e.g., What is the likelihood? How likely is it? What is the probability?) The formula we present is used for determining the probability of an event if all outcomes are equally likely.

Problem 1

There are 75 possible numbers in bingo. Find the probability that the first number selected is odd.

① *Number of favorable outcomes*
There are 38 odd numbers that are less than or equal to 75.

② *Number of possible outcomes*
There are 75 numbers in bingo.

③ *The probability of getting an odd number in bingo is*
$P = \dfrac{38}{75}$

Problem 2

A coin is tossed 3 times. Find the probability of tossing at least one heads.

① *Number of favorable outcomes*
There is only one way we would not toss a heads and that is if we toss three tails. So, we subtract the 1 from the number of possible outcomes $8 - 1 = 7$ favorable outcomes.

② *Number of possible outcomes*
There are $2 \times 2 \times 2 = 8$ possible outcomes.

The probability of getting one heads is:

$P = \dfrac{7}{8}$

Uniform Motion

$r \times t = d$ (r = rate, t = time, and d = distance)

A common form of word problem in algebra is the "uniform motion" problem. These word problems ask students to read a travel question and apply a formula to determine some aspect of the question (e.g., How long will it take? Who will get there first? Where will they meet halfway?).

Problem 1

Frank and Kim leave their cabin at 8:00 AM and paddle their kayaks down the river at the constant rate of 8 kilometers per hour. At 11:00 AM, Juan leaves the cabin in a motorboat and travels down the river with four friends in the boat. If the motorboat travels at 20 kilometers per hour, how long before Juan catches up with the kayaks?

Frank and Kim	Juan
rate = 20 kmph	rate = 20 kmph
time = $t + 3$ (the number of hours Frank and Kim left before Juan)	time = t
	distance = d

$8 \times (t + 3) = 20t$

$8t + 24 = 20t$

$24 = 12t$

$\dfrac{24}{12} = \dfrac{12t}{12}$

$t = 2$ hours or 1:00 PM before Juan catches up with the Kayaks

Problem 2

A mover left Lynchburg at 8:00 AM and drove to Asbury Park, off Exit 8A of the Turnpike, at 55 mph. At 9:00 that morning, the five members of the Jones family left Lynchburg on the same route and arrived in Asbury Park at the same time as the mover (4:00 PM). What was the Jones' average speed?

Mover	Jones Family
rate = 55 mph	rate = r
time = 8 hours	time = 7
distance = d	distance = d

$55 \times 8 = r \times 7$

$440 = 7r$

$\dfrac{440}{7} = \dfrac{7r}{7}$

62.86 mph was the average speed of the Jones family

Compound Interest

$$M = P(1 + i)^n$$

(M = final amount, P = principal amount, and i = interest rate per year, and n = number of years)

Another formula used often in problem solving is calculating compound interest, interest that is paid on both the principal and on any interest from past years. Common questions that are asked with these types of problems include: How much will I earn? How much will I have to pay back? How much more will I earn with compound over simple interest?

Problem 1

How much money will you earn if you invest $1,000 for 5 years at 5% compound interest (compounded annually)? Round to the nearest whole dollar.

$P = \$1,000$

$i = 5\%$

$n = 5$

$M = \$1,000(1 + .05)^5$

$M = \$1,000(1.05)^5$

$M = \$1,000(1.28)$

$M = \$1,276$

Problem 2

You need a loan for $2,000 and have narrowed the lender to two banks. Bank A will loan you the $2,000 for 4 years at 8% compound interest, and Bank B will loan you the $2,000 for 3 years at 10% compound interest (compounded annually). At which bank will you pay back less money?

Bank A	Bank B
$P = \$2,000$	$P = \$2,000$
$i = 8\%$	$i = 10\%$
$n = 5$	$n = 3$
$M = \$2,000(1 + .08)^4$	$M = \$2,000(1 + .10)^3$
$M = \$2,000(1.08)^4$	$M = \$2,000(1.10)^3$
$M = \$2,000(1.36)$	$M = \$2,000(1.33)$
$M = \$2,721$	$M = \$2,662$

You will pay back less money at Bank B.

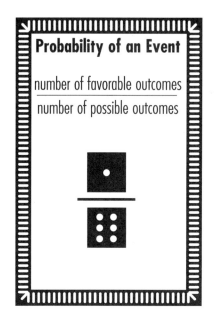
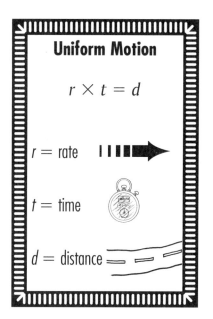
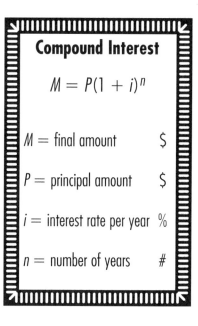
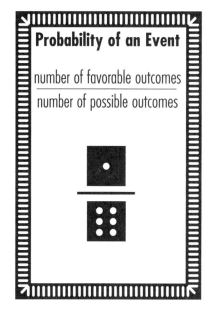
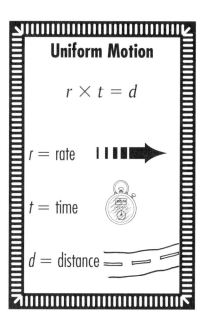
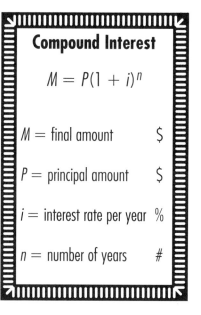
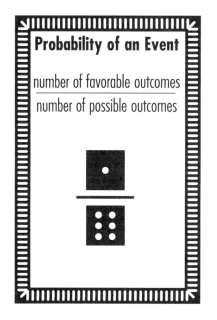
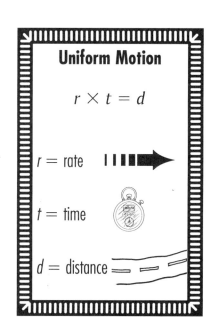
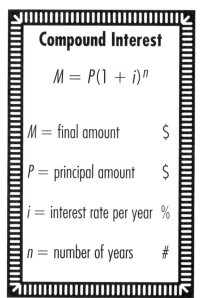

Idea 30
Maze It or Cloze It

To be competent math problem solvers, students must know what algorithm or procedure to use to solve problems. The maze or cloze procedure can be used to help students choose an action verb that will help them decide which procedure they need to complete the math problem correctly. In the maze procedure, students choose one word from several words to make a problem true. In the cloze procedure, students insert a word in spaces where text has been left out. Examples using addition and subtraction with regrouping are provided. Note that the Maze It example has two formats for the word choices (i.e., vertical and horizontal).

Here's how Maze It works.

❶ Develop a set of math problems that can be solved using one of several mathematical procedures.

❷ Provide a choice of three action verbs that indicates how to solve the problem.

❸ Have students circle the action verb that makes the problem correct and then solve the problem to check their answers.

Here's how Cloze It works.

❶ Develop a set of problems that can be solved using one of several algorithms.

❷ Leave a blank where an action verb should go that indicates how to solve the problem.

❸ Have students write in an action verb that makes the problem correct and then solve the problem to check their answers.

Note. This idea is based on *Mathematics for the Mildly Handicapped: A Guide to Curriculum and Instruction* (pp. 179–180), by J. F. Cawley, A. M. Fitzmaurice-Hayes, and R. A. Shaw, 1988, Boston: Allyn & Bacon. Copyright 1988 by Allyn & Bacon.

Maze It

Circle the word that makes each problem true. Next, solve the problem in the Check It column to check your answer.

Problems	Check It
❶ Nirumba had $2.25. He { spent / borrowed / made } $1.35. Now he has $0.90.	
❷ Chloe had 296 pennies in her penny jar. She { found / spent / loaned } 78 pennies. Now she has 374 pennies.	
❸ Kevin had 42 grapes. He { ate picked bought } 16 grapes. Now he has 26 grapes.	
❹ Maria had 168 sports cards. She { swapped bought sold } 44 sports cards. Now she has 212 sports cards.	

© 2006 by PRO-ED, Inc. Idea 30

Cloze It

Write a word in the blank that makes each problem true. Next, solve the problem in the Check It column to check your answer.

Problems	Check It
❶ Jonathon had $3.46. He _____ $1.35. Now he has $4.81.	
❷ Shanelle had 282 marbles. She _____ 96 marbles. Now she has 186 marbles.	
❸ Pablo had 152 stamps in his stamp collection. He _____ 83 stamps. Now he has 69 stamps.	
❹ Danielle had 28 meters of ribbon. She _____ 17 meters. Now she has 45 meters of ribbon.	

© 2006 by PRO-ED, Inc. Idea 30

Maze It

Circle the word that makes each problem true. Next, solve the problem in the Check It column to check your answer.

Problems	Check It
❶	
❷	
❸	
❹	

Idea 30

Cloze It

Write a word in the blank that makes each problem true. Next, solve the problem in the Check It column to check your answer.

Problems	Check It
❶	
❷	
❸	
❹	

Idea 30

Idea 31
Peer Partners

Peer Partners provides steps for students to follow when placed in dyads or small groups. Ideas 32 and 33, Assigning Students to Groups and Collaborative Problem Solving, provide methods for assigning students to various types groups and improving students' ability to work in groups.

Here's what to do.

❶ Assign students to dyads or groups, and provide one problem with an answer key to the group.

❷ Give each group a copy of the Peer Partners form, and explain the steps using a sample problem.

❸ Once students are comfortable with the process, give them several problems to complete.

Tip:

Post the sign-up sheet provided at the end of this idea for students to make appointments with the teacher for assistance with problems. The students will write the problem number in the first column and their names in the second column. The teacher places a checkmark in the last column after he or she finishes meeting with the group.

Peer Partners

1. Solve the problems by yourself.
2. Check the solutions with the scoring key.
3. If one group member makes an error, other group members show how to solve the problem.
4. If all group members make an error, the group gets help from the teacher.

Peer Partners

1. Solve the problems by yourself.
2. Check the solutions with the scoring key.
3. If one group member makes an error, other group members show how to solve the problem.
4. If all group members make an error, the group gets help from the teacher.

Sign Up for Teacher Talk Time

Problem Number	Group	✔

Idea 32
Assigning Students to Groups

Much has been written on using collaborative group work in math classes. In fact, it is encouraged by the National Council of Teachers of Mathematics (NCTM, 2000). Students enjoy group work because it provides opportunities for both social and cognitive collaboration. Sometimes you want to assign students to groups based on achievement level, work style, or diversity of skills and backgrounds, but there are other times when randomly assigned groups are appropriate. Random work groups allow students to get to know each other and provide opportunities for a variety of learners to interact. This idea provides two processes for randomly assigning students to groups.

Method 1: Rolling Groups

This method should ensure a random selection and is fun for the students. It requires the Rolling Groups Record Form (provided) and a die.

❶ First, put the Rolling Groups Record Form on the overhead, or copy it onto the board or a sheet of poster paper. There are two forms, one that includes six groups with five students in each group and one with five groups of four students each. You can choose either form, or you can reduce the slots for smaller classes by crossing off unneeded groups or student slots.

❷ Next, give the first student in the class the die and ask him or her to roll. The number rolled is the number of his or her assigned group, so write that student's name on the line of the group with the same number.

❸ Ask the first student to pass the die to the next student and repeat until everyone in the class has had a chance to roll and is assigned to a group. If a group is filled (e.g., all the number 4 slots already have names in them), ask the student to roll again.

After students have been randomly assigned to a Rolling Group, you can further clarify their roles within the group by assigning responsibilities based on their position on the form. For example, all students in the A column will gather and distribute materials, the Bs will be recorders, Cs will be managers, Ds will be time keepers, and Es will be presenters.

Note. Rolling Groups and Time for Group Work are from *Practical Ideas That Really Work for English Language Learners* (pp. 11–16), by K. McConnell, D. Campos, and G. R. Ryser, 2006, Austin, TX: PRO-ED. Copyright 2006 by PRO-ED, Inc. Reprinted with permission.

Method 2: Time for Group Work

This method provides a way of assigning students to pairs and ensures that the students have an opportunity to work with many other different students. All you need is the clock template that we have provided.

❶ Photocopy enough clocks for your class. Before distributing, write the names of students in the classroom for each hour on the clock. Consider writing the names in alphabetical order or skipping every second or third name (names will repeat). Make sure that no student is on the same time for more than one clock.

❷ When it comes time for group work, say "I want you to work with your 3 o'clock partners" or "Find your 8 o'clock partners."

❸ After using the first clock for a week or two, pass out a new clock, making sure that each student receives a different set of names.

Rolling Groups Record Form

Group Number	Group Members			
	A	B	C	D
1				
2				
3				
4				
5				

Rolling Groups Record Form

Group Number	Group Members				
	A	B	C	D	E
1					
2					
3					
4					
5					
6					

Time for Group Work

Directions: Write the name of one student for each hour on the clock.

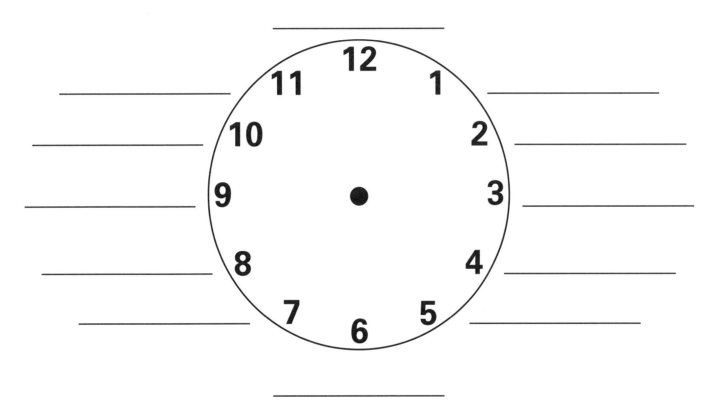

Idea 32

Idea 33
Collaborative Problem Solving

Recent research shows that the effects of collaborative group work in math problem solving can often but not always have positive outcomes on cognitive processes. Collaborative groups typically are more effective with open-ended math problems. In addition, teachers should model strategies that are likely to lead to successful outcomes, including listening skills, resolving disagreements, and reviewing progress. This idea presents two collaborative group processes that can be effective for improving students' math problem solving and improving their ability to work in groups effectively.

Method 1: Jigsaw Groups

The jigsaw was developed by Dr. Elliot Aronson, professor emeritus at the University of California in Santa Cruz. Jigsaw groups work very well when students are studying a particular math problem-solving strategy, such as Test and Guess, or a particular math formula, which lends itself to problem solving, such as *rate × time = distance*. The jigsaw technique described below is a strategy that forces all students to become responsible for their own learning.

❶ Divide the class into small groups of four or five students, depending on the number of problems to be solved.

❷ Each member of each group will be given one problem in which to become an expert. With five groups, there will be five individuals, one from each group, studying the same problem. Use the Jigsaw Topics form to assign problems by number.

❸ Allow time for students to solve or attempt to solve their problem. This works best when you have already introduced the type of strategy or formula you will use to solve the problems.

Tip:
Make four copies of the four-piece puzzle or five copies of the five-piece puzzle, depending on the number of groups you have. Give each group member one piece. This number corresponds to the problem on the Jigsaw Topic page.

Note. Create Jigsaw Groups is from *Practical Ideas That Really Work for Students Who Are Gifted* (pp. 103–107), by G. R. Ryser and K. McConnell, 2003, Austin, TX: PRO-ED. Copyright 2003 by PRO-ED, Inc. Reprinted with permission.

❹ After an adequate amount of time, all students studying the same problem get together and discuss the process used to solve it. Your role is to make sure that groups are functioning well together and that nobody is dominating the discussion or being disruptive. With time, you will want the group members to become more responsible for handling these situations.

❺ Have all group members reconvene in their original groups. Each expert presents his or her problem to the rest of the group members. Group members are encouraged to ask questions to clarify the problem solving process.

Method 2: Heads Together

This method provides a way of reviewing information that has been previously presented. It works well with questions that allow students to easily come to consensus and is a great way to review for a test.

❶ Develop several review questions.

❷ Divide students into groups of three to five students. Have the students in the group number off.

❸ Ask a question and have the groups consult quietly to determine the answer.

❹ Call a number and the students with that number raise their hands to respond. The teacher can call on one student or have the students respond together.

❺ Have the rest of the students agree or disagree using a signal, such as thumbs up or thumbs down.

Jigsaw Topics

1. _____

2. _____

3. _____

4. _____

Jigsaw Topics

1. _____

2. _____

3. _____

4. _____

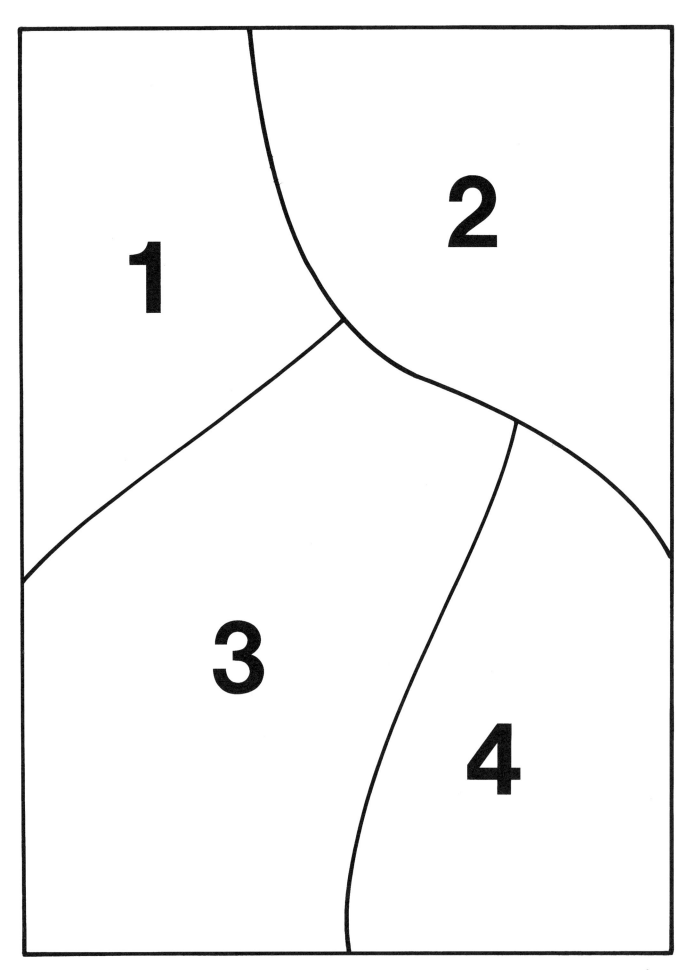

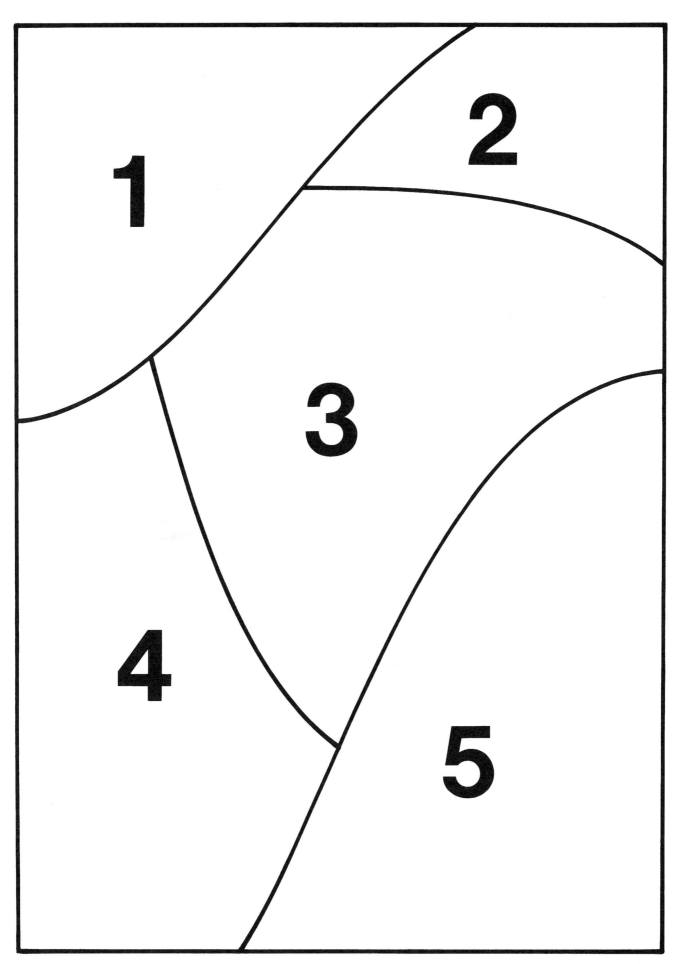

Idea 34
Use a Matrix

We are all familiar with the sequence of skills that is associated with computation in math. Students first learn to count and then typically learn to add (single digits and later with regrouping), subtract, multiply, divide, and so on. For math problems, a similar approach can be used as students gradually learn to read and respond to more challenging problems. This process can be facilitated through the use of a math problem-solving matrix. Matrices can be developed from a bank of written math problems for any level, across any grade. This strategy assists teachers in structuring a math problem-solving curriculum so that students can progress from simple to more complex problems.

Another reason for using a matrix is to teach students how to apply their knowledge of how to solve one type of math problem to solving novel problems. To help students increase their awareness that novel problems are related to previously solved problems, change features in problems gradually, and teach students to search for connections to familiar problem structures. For information on teaching transfer, see Fuchs, Hamlett, and Appleton (2003).

We decided to include this idea, even though it requires a substantial investment of time, because the underlying concept is powerful and useful and very much in line with the thinking of John Cawley, whom we respect greatly for his work in the area of mathematics and students with special needs. We remain hopeful that some users of this book will implement this idea.

Here is how to develop a matrix.

❶ Select some word problems from a basal program or math curriculum.

❷ Develop a word problem-solving matrix that reflects your math curricular objectives by writing the details of how the set of problems will change as the row headers and the operations used as the column headers (two example matrices are provided at the end of this idea).

❸ Complete the matrix by writing the page numbers and item numbers in the cell that matches that type of problem.

❹ Teach students to do the first set of problems found in column 1, row 1. Once students have demonstrated mastery of the first set of problems, present the problems found in column 2, row 1. Continue throughout an instructional unit or other time period so that students are gradually challenged by more difficult word problems.

Typical ways in which matrices can reflect increased difficulty include the arithmetical operation selected, the reading level of the problem, the absence or presence of distracters, one-step or multiple-step operations, the structure of the problem, and the magnitude of the number or numerals within the problem. An example of a graduated problem sequence (from easy to difficult) includes problems

- with single words or phrases.
- with sentences, numbers aligned vertically.
- in paragraph form.
- without extraneous information.
- with extraneous information.
- created by students.

Use a Matrix

	Addition, no extraneous information	Addition, with extraneous information	Subtraction, no extraneous information	Subtraction, with extraneous information
1 Digit 2 Items	p. 32, #4 Maria earned $4 raking her neighbor's lawn. She earned $7 last week. How much does Maria have now?	p. 43, #8 Derrick made 2 batches of cookies. He wanted to share them with his neighbors. The next day he made 4 more batches of cookies. How many batches of cookies did he make in all?		
1 Digit 3 Items	p. 32, #12 Sandra had 8 quarters. Her mom gave her 3 more. Later that week she visited her grandfather who gave her 5 more quarters. How many does she have altogether?	p. 45, #6 Jerome has 5 fish in his fish tank. He likes the angelfish the best. He bought 2 more fish on Saturday. On Tuesday, his aunt gave him 4 more fish. How many fish does Jerome have altogether?		
2 Digits 2 Items				
2 Digits 3 Items				
Varied				

Use a Matrix

	Addition, no extraneous information	Addition, with extraneous information	Subtraction, no extraneous information	Subtraction, with extraneous information
1 Digit 2 Items				
1 Digit 3 Items				
2 Digits 2 Items				
2 Digits 3 Items				
Varied				

Use a Matrix

	Addition or Subtraction, no unknowns	Addition or Subtraction, one or more unknowns	Multiplication or Division, no unknowns	Multiplication or Division, one or more unknowns
Fractions, same denominator				
Fractions, different denominator				
Mixed numbers				
Proportions				
Varied				

Use a Matrix